Complete Kenya Cookery

Ruth K. Oniang'o
Asenath J. Sigot

Arnold Hodder Africa

© Ruth K. Oniang'o and Asenath J. Sigot 1987

First published in Great Britain 1987 by
Edward Arnold (Publishers) Ltd, 41 Bedford Square, London WC1B 3DQ

Edward Arnold (Australia) Pty Ltd, 80 Waverley Road, Caulfield East,
  Victoria 3145, Australia

Edward Arnold, 3 East Read Street, Baltimore, Maryland 21201, USA

All rights reserved. No part of this publication may be reproduced, stored
in a retrieval system, or transmitted in any form or by any means, electronic,
photocopying, recording, or otherwise, without the prior permission of
Edward Arnold (Publishers) Ltd.

Text set in 10/12 pt Century Schoolbook
by Colset Private Limited, Singapore

Printed by
General Printers Ltd. Homa Bay Road,
P.O. Box 18001, Nairobi.

Contents

Acknowledgements	v
Introduction	vii
Nutrition	1
Food hygiene	4
Food storage	5
Tips for successful cooking	6
Planning and serving family meals	7
Breakfast	9
Light dishes and snacks	14
Soup	23
Eggs	27
One-pot dishes	33
Stews	42
Meat	54
Offal	64
Poultry	68
Fish	74
Cereal-based dishes	81
Pulses	94
Tubers	107
Other vegetables	115
Sauces	126
Bread	131
Cakes and biscuits	134

Desserts, sweets and puddings	149
Beverages	157
Recipes from other countries	161
Glossary of cooking terms	172
Glossary of food terms	175
References	177
Index	179

Acknowledgements

We would like to take this opportunity to thank our students, individually and collectively, for the effort they put into this project and for the team spirit that made it possible.

We would also like to thank Charles Ogutu Ohomo and the members of staff of the Home Economics Department at Kenyatta University for their support and contributions. Special appreciation goes to those individuals who, though not members of the Department, were willing to share their recipes with us. We thank all who contributed in the slightest way.

Introduction

Complete Kenya Cookery has been developed to suit the needs of students following the cookery component of the Secondary Home economics syllabus.

The collection of recipes in this book is a result of the efforts of students of the Experimental Foods course at the Department of Home Economics, Kenyatta University, Nairobi. Many of the students involved were experienced teachers who had taught Home Science in schools and had come to pursue a degree course in Home Economics. The rest of the students were high school graduates whose fresh ideas blended well with those of the more experienced group. The Experimental Foods course, whose main objective is to acquaint students with locally available foodstuffs and to give them an opportunity to experiment, formulate new recipes and improve on existing recipes, is a third-year course at the University. This is when students put their understanding of the science of food preparation into practice. Teaching members of the Department of Home Economics were very supportive of this project and contributed some of their own favourite recipes for inclusion.

Recipes were chosen from many different areas of the country and show the range of dishes and foodstuffs available. There are also recipes from other countries such as Ghana, Liberia and Guyana which were provided by members of faculty of Kenyatta University.

Each recipe was prepared three times. The first preparation was the most experimental: quantities of ingredients were recorded, methods of preparation noted and a draft recipe written. The dish was then evaluated. Comments at the first evaluation were considered during the second trial and there was a third and final preparation. Objective evaluation was made by students themselves for other students.

We welcome any constructive comments from our readers to help us improve future editions. If you have any interesting recipes you would like to share, please send them to us.

Ruth Oniang'o	Asenath Sigot
Box 43888	Box 43844
Nairobi	Nairobi

Nutrition

The science of nutrition is concerned with what happens to the food we eat from the time we ingest it, through digestion, absorption, utilisation and excretion. What we eat is dictated by appetite, which is psychological, and hunger, which is physiological.

What we eat depends on a number of factors, the main one being what is available. The food we eat must be in a state which can be consumed in amounts adequate to our physiological needs. This is where food preparation comes in. The food we prepare is no use if it cannot be enjoyed and therefore eaten in sufficient amounts.

Food provides us with the nutrients that the body requires for proper maintenance. We need carbohydrates, proteins, fats, vitamins, minerals, fibre and water. A balanced diet, containing all these in the proper proportions, is essential for good health.

Carbohydrates

These are usually considered the major nutrient because they can avert starvation. The primary function of carbohydrates is to provide us with energy to maintain body temperature and to utilise other nutrients.

Proteins

Proteins promote growth and are hence crucial in the diets of youngsters who are still growing. They are also necessary for the synthesis of hormones and enzymes in the body and they serve as secondary sources of energy. If the diet does not contain enough carbohydrates, proteins can be broken down to provide energy, but this means that they cannot perform other functions.

The majority of people in developing countries such as Kenya derive both carbohydrates and proteins from plant sources such as cereals, tubers, roots and legumes.

Fats

Fat is a source of energy. It also serves the function of insulating the body. Some organs, such as the kidneys, are protected by fat. Animal fats contain vitamins A and D. Fats also have a high satiety value. Apart from commercial fats used in food preparation, Kenyans also obtain fat from such food as milk, groundnuts, simsim, fatty meat, coconuts and avocados.

Vitamins

Vitamins are required by the body in minute amounts, but their functions are vital. The main vitamins are:

Vitamin B Complex

This group of vitamins plays a major role in various metabolic processes. It includes vitamins B_1 (thiamine), B_2 (riboflavin) and niacin. A mixed diet including whole grain cereals, meats, legumes and dark green leafy vegetables will give adequate amounts of this group of vitamins.

Vitamin C (Ascorbic Acid)

Protects against infections and maintains healthy gums. Bleeding gums are a symptom of vitamin C deficiency. Fresh vegetables, if not exposed to sunlight or over-cooked, provide us with this vitamin. The best sources, however, are citrus fruits (oranges, lemons, limes, grapefruit), guavas and pawpaws.

Both Vitamins B and C are water-soluble and can be lost into cooking water. Vitamin C is easily lost by exposure to air and heat.

Vitamin D (Calciferol)

Manufactured under the skin by ultraviolet light from the sun. Growing children as well as adults should be exposed to plenty of direct sunlight. Vitamin D works with calcium and phosphorus to form strong teeth and bones and thus prevents rickets. Margarine is fortified with this vitamin and fish and cod-liver oil are the other food sources.

Minerals

Like vitamins, minerals are required in small amounts and are essential for the maintenance of body processes. The best known are:

Iron

Needed for blood formation. Iron deficiency results in anaemia. Dried legumes, green leafy vegetables, dried fruit, liver, blood and molasses are good sources.

Calcium and Phosphorus

With vitamin D, calcium and phosphorus are responsible for healthy bone development. Calcium is also needed for the clotting of blood. Calcium is present in dairy products such as milk and cheese, in eggs, fish bones, dried legumes and most proteins.

Fluoride

Occurs in very high concentrations in some areas of Kenya, where it can cause mottled teeth. Ingested in the right amounts, though, fluoride is necessary for the prevention of dental cavities. It occurs naturally in water, concentrations varying from area to area. Fluoridated toothpaste is available in Kenya and the non-fluoridated type has more recently become available.

Iodine

The concentration of iodine in soils varies from area to area and this affects the levels of the mineral in foods. There are areas of Kenya where the iodine content of the soil is very low. In these areas, there tends to be a high incidence of goitre (enlargement of the thyroid gland). Since the problem was identified, iodised salt has become available on the Kenyan market. Iodine facilitates the production of thyroxine, a hormone required for metabolism by the thyroid gland. When iodine is not available in adequate amounts, the thyroid gland enlarges as it tries to produce enough thyroxine for the body. Iodised salt is the best source of iodine.

Water

Water is essential for life. It is needed in blood formation, as the medium for transporting nutrients and other substances to various parts of the body, and as a carrier of waste products from the body.

Fibre

Fibre, though not usually considered a nutrient, is a very important component of our diet. Fibre, also known as roughage, is the rough material found mostly in fruit, vegetables and whole cereals. Its function is to aid digestion and elimination. There is some evidence to suggest that eating adequate amounts of fibre plays a role in the prevention of some of the cancers of the gastro-intestinal tract.

Food hygiene

Food spoilage and food-borne diseases are caused by bacteria and other micro-organisms. The food we eat, if we are not careful, can be harmful. A warm climate such as that of Kenya provides just the right environment for many micro-organisms to thrive. Following a few simple rules will ensure that food is safe to eat.

1. After buying perishable food, take it home straight away for immediate use or storage. Meat, poultry and fish should be kept separately in covered dishes in the refrigerator so that bacteria cannot spread.
2. Keep hot food *hot* (above 140 °F or 60°C) and cold food *cold* (below 40 °F or 4°C) to prevent bacteria growing. If hot food is prepared in advance, it should be kept between 140 °F and 150 °F until it is eaten.
3. Do not let cooked food stand at room temperature for more than two or three hours including preparation, storage and serving time.
4. Hands should always be clean when you handle any food. After handling raw meat, fish, poultry or eggs, wash your hands with soap and water before working with other foods.
5. Never place other food on a surface where you have had raw meat or poultry until you have thoroughly cleaned it. Dishes, knives and any other utensils used should also be thoroughly cleaned. This prevents bacteria spreading to cooked food and salads.
6. Clean chopping boards by scrubbing with soap and hot water, rinsing thoroughly and then applying a suitable disinfectant as directed on the label.
7. Quickly refrigerate cream, custard and any food with a cream filling. It is dangerous to let food of this kind stand at room temperature.
8. Poultry, pork and pork products must be thoroughly cooked. The heat will destroy the micro-organisms in the meat.
9. Wash eggs before using them because they carry bacteria on their shells. Wash hands thoroughly after handling raw eggs. Do not, however, wash eggs before storage, however dirty they may be, since this facilitates the entry of bacteria into the sterile inside through the porous shell.
10. Thaw frozen poultry in the refrigerator. If you must thaw it more quickly, put it in a waterproof polythene bag, seal it tightly and put it in cold water. Thawing frozen poultry at room temperature gives bacteria a chance to grow.
11. Dishes with meat fillings should not be left at room temperature for more than an hour.
12. Washing hands after visiting the toilet, scratching, touching hair, touching animals, sneezing or handling different foods should become a habit.

Food storage

Wastage through food spoilage because of poor storage or poor planning can be avoided. Storing food is easier if a refrigerator is available. Without refrigeration facilities, certain principles must be observed:

1. Green leafy vegetables, meat and dairy products do not keep well without refrigeration, so buy only what can be used immediately. Withered fruit and vegetables lose their taste and nutritive value.
2. Meat can be preserved by salting and drying it on charcoal or under the grill.
3. Tuberous and root crop products should be kept as dry as possible, for example by storing them in sand. Cassavas can be stored immersed in cold water for a day or two or kept underground in the soil. Exposure to air makes them spoil very quickly.
4. Cereals should be kept as dry as possible to prevent dangerous mould developing. Grain or flour that has moulded should not be consumed. Keep cereals and nuts in the same way.
5. To freeze vegetables such as sukumawiki, clean them thoroughly, chop finely, immerse in boiling water for about 3 minutes, drain, cool, wrap in a polythene bag and freeze. Most frozen food will keep for three months or longer.
6. Eggs will keep in a refrigerator for as long as two months, but they go bad within two weeks at room temperature.
7. Keep meat and any meat dishes in the coldest part of the refrigerator.
8. Fats such as margarine and kimbo do not need to be refrigerated because they contain preservatives. The amounts bought for household use are usually finished before they can go bad.
9. Sodium bicarbonate absorbs strong odours. Use it to clean and remove strong flavours from cooking utensils and put a container of sodium bicarbonate in the refrigerator.

Tips for successful cooking

1. Always have some kind of cooking acid on hand, such as vinegar, lemon juice, tartaric acid or citric acid. These acids can be used as flavour enhancers in both sweet and savoury dishes and to prevent certain fruit and vegetables, such as bananas, pears and avocados, going brown.
2. An individual's palate gets used to a certain level of salt. Salt is a flavour enhancer, even in sweet dishes, if used with discretion. Salt is best added to dishes at the beginning of the cooking period, except for meat, which it tends to toughen if added too early. However, try to accustom your family to minimum levels of salt in their food. A diet containing too much salt has been linked to the incidence of high blood pressure.
3. Fat is not only a source of energy; it also enhances the flavour of foods. Fat should be used sparingly because it can make food greasy and unpalatable. Margarine, vegetable fat (such as kimbo), oil, ghee, butter, groundnut and simsim pastes can all be used in cooking. Fats are also good at absorbing and retaining flavours. This means that fat or dishes containing fat should not be stored near strongly flavoured food.
4. Mushrooms and groundnuts are very good flavour enhancers and they do not need to be used in large quantities. A number of flavouring agents now on the market can be successfully used to improve bland dishes. Mchuzi mix and Grevi mix are two examples.
5. Milk may be fresh, sour (mala) or powdered. Curdled milk need not be thrown away; it can be used in some vegetable dishes, porridge or baked products. If, however, the milk has soured, its acid flavour can be counteracted by adding $\frac{1}{2}$ tsp bicarbonate of soda to each cup of sour milk.
6. The cooking time of dried legumes can be shortened by first soaking them in cold water for 8–12 hours, in warm water for 3–4 hours, or in hot water for 1–3 hours. The soaking water can sometimes be used, but overnight soaking water should be thrown out, since it usually has a bad odour and can affect the flavour of the end product.
7. Pressure-cooking saves fuel, time and money, but some flavour can be lost because the food does not cook long enough to develop its full flavour. This applies particularly to legumes and meat.
8. Water drained from such foods as vegetables, irio or potatoes should not be thrown out, because it contains valuable nutrients. Drained water should be used in mashing or in sauces and stews.
9. To condition a frying pan to which food sticks, rub it thoroughly, first with dry salt, then with oil. Do not use the salt for cooking.
10. Rinsing onions in warm water will help to prevent sliminess, weaken the smell and help to retain their crispness.

Planning and serving family meals

Family meals must be carefully planned and served. A knowledge of nutrition and health must be applied to every meal consumed. Meals should:

1 be nutritionally adequate
2 suit the family
3 be possible to prepare
4 satisfy hunger
5 be pleasing and appetising.

A balanced diet means using a wide variety of foods; it does not mean eating the same type of food day after day.

Menu planning

Planning the meals for a week is useful as a guide even though the food suggested for one meal may be substituted at another. Working out a meal pattern helps prevent duplicating the same kind of meal day after day.

Points to consider in menu planning

1. A balanced diet provides sufficient carbohydrate, protein, fat, minerals and vitamins for all the family.
2. Individual needs: palatability, economy and ease of preparation. Consider the number of people in the family, their ages, health, occupations, activities, likes and dislikes. Meals eaten away from home should also be considered when planning meals for the day.
3. Variety in kinds of food, flavours, textures, colours, temperature and methods of preparation.
4. Planning ahead. Planning menus in advance saves time, energy and money and ensures nutritious, attractive and flavourful meals. Plans should be flexible enough to allow for adjustment if necessary.
5. Serving meals. Hot food should be served quickly, while it is still hot. Garnish it to make it look tempting and set the table so that it looks inviting.

Suggestions for a day's meals

MENU 1	MENU 2
Breakfast Wimbi (millet) porridge with lemon or honey A ripe banana or 1 slice of bread 1 boiled egg	*Breakfast* Scrambled egg on bread Boiled sweet potato and tea with milk
Lunch Ugali Sukumawiki (kale) with groundnuts and carrot Fresh fruit or Chicken curry with rice and green vegetables with tomatoes	*Lunch* Boiled rice and meat stew with green vegetables or Mashed bananas with groundnut sauce, green peas and tomatoes
Supper Chapati with rest of the chicken or with beef stew or vegetable soup Tea or coffee	*Supper* Spaghetti with meat sauce and green vegetables Tea or coffee

Breakfast

Breakfast is an important meal because it provides the energy we need to start the day. Cereals are the most common constituent of breakfast dishes. Those widely available in Kenya include maize, millet, sorghum and wheat, and dishes based on these are part of many Kenyans' staple diet. Porridge is a very popular breakfast dish, and is most commonly based on maizemeal or millet flour: it can be enriched with fat and milk. Porridge is also used for weaning infants. Other foods rich in carbohydrates, such as tuberous and root crops, are also suitable for breakfast.

Nyuk Abak Bel
(Fermented Sorghum Porridge)

Pauline Owiti

Serves 4

Ingredients

½ cup red sorghum flour
½ cup unga baridi (very fine white flour)
5 cups water
2 tbsps sugar

Method

1 Mix both types of flour with 1 cup of warm water in a non-corrosive container. Cover and leave it to ferment in a warm place for two days.
2 Boil 3 cups of water. Blend the flour mixture gradually into the boiling water, stirring well to prevent it becoming lumpy.
3 Add 1 cup of water, still stirring. Add the sugar and continue simmering for 20–30 minutes or until the foam begins to disappear. Serve hot.

Assorted Flour Porridge
Zipporah Kabugu

Serves 4

Ingredients

6 cups water
½ cup millet (wimbi) flour
¼ cup sorghum flour
¼ cup maizemeal
½ cup milk
sugar to taste

Method

1 Boil 5½ cups of water. Mix the flour with ½ cup of cold water; stir until there are no lumps.
2 Add the mixture to the boiling water, stirring continuously.
3 Lower the heat and simmer for 30 minutes, stirring continuously.
4 Add the milk and bring to the boil.
5 Remove from the heat and add sugar to taste.
Variation: lemon juice may be added just before serving.

Wimbi Porridge
Ruth Mititi

Serves 2

Ingredients

4 cups water
½ cup wimbi flour
1 cup fresh milk
2 tbsps sugar
4 tbsps (⅓ cup) lemon juice

Method

1 Boil 3 cups of water. Mix the flour with 1 cup of water; stir until there are no lumps.
2 Add the mixture to the boiling water, stirring continuously. Simmer for 5 minutes.
3 Add the milk and simmer for a further 10 minutes.
4 Add the lemon juice and sugar, stirring well, and simmer for 2 minutes. Serve hot.

Wimbi Porridge (with sour milk) *Rachel Jondiko*

Serves 2

Ingredients

2 cups water
½ cup wimbi flour
2 tbsps sugar
2 cups sour milk

Method

1. Boil 1½ cups of water. Add wimbi flour paste, made with ½ cup of cold water, and keep stirring until it thickens and boils.
2. Add the sugar, then the sour milk, and boil for 2–3 minutes. Serve hot.

Fermented Porridge *Pauline Githiru*

Serves 4

Ingredients

½ cup maize flour
¼ cup wimbi flour
¼ cup sorghum flour
7 cups water *or* 4 cups water and 3 cups yoghurt or sour milk
sugar to taste

Method

1. Mix all the flour with 4 cups of water in a non-corrosive container. Cover and leave to ferment for two days.
2. Sieve the liquid into a saucepan and add 3 cups of fresh water, yoghurt or sour milk.
3. Bring this liquid to the boil and add the fermented paste, stirring well to prevent lumps.
4. Simmer for 15–20 minutes, adding more water if necessary, until a smooth, slightly runny consistency is reached.
5. Remove from the heat and add sugar to taste. Allow to cool slightly and serve in glasses or mugs.

Note: traditionally, fermented porridge was served in half-calabashes.

Fermented Sorghum and Wimbi Porridge

Agnes Masika

Serves 4

Ingredients

½ cup sorghum flour
½ cup millet flour
10 cups water
sugar to taste

Method

1. Mix the flour with $1\frac{2}{3}$ cups of water in a non-corrosive container. Cover and leave in a warm place for three days.
2. Mix fermented flour with $1\frac{2}{3}$ cups water and add it to $6\frac{2}{3}$ cups boiling water, stirring continuously until it thickens. Boil for 5 minutes. Add sugar to taste and serve hot.

Green Maize Porridge

Eva Gatere

Serves 2

Ingredients

$3\frac{1}{2}$ cups raw green maize
3 cups water
½ cup sugar

Method

1. Pound the maize until soft, and pass it through a sieve. Pour the water through and collect all the liquid.
2. Pound the maize again and repeat the process, using the water collected in step 1.
3. Store this water, covered, for 3–4 days until fermented.
4. Boil 3 cups of fresh water and stir in the fermented liquid.
5. Stir until it starts boiling, then lower the heat and simmer for about 15 minutes. Add the sugar and serve.

Breakfast 13

Ruguru Porridge

Sylvia Murungi

Serves 4

Ingredients

1 bunch arrowroot leaves
7 cups water
1 tbsp baking soda
1 cup wimbi flour
juice of 1 lemon
$\frac{1}{4}$ cup sugar

Method

1. Remove and discard the veins of the arrowroot leaves, wash the leaves and shred them finely.
2. Put 6 cups of water and the shredded leaves into a medium-sized saucepan. Add the baking soda and boil until the leaves are tender.
3. Whisk until the leaves are completely dissolved in water; if the leaves do not dissolve completely, pass the solution through a sieve.
4. Measure the solution into another pan and add enough water (about 1 cup) to make $1\frac{1}{4}$ litres.
5. Stir in the flour and bring to the boil, stirring constantly. Boil for 10 minutes.
6. Add the lemon juice and sugar. Stir well and serve hot.

Note: the baking soda enhances the greenish colour of the porridge.

Yam Porridge

Lois Munoru

Serves 2

Ingredients

1 medium yam
pinch of salt
1 cup warm milk
sugar to taste

Method

1. Wash the yam and boil until tender. Peel and grate it.
2. Add a pinch of salt, then mash with a wooden spoon, adding a little warm milk at a time, until the mixture is of the desired consistency.
3. Add sugar to taste and serve hot or cold.

Light dishes and snacks

Snacks are very light meals that are usually simple and quick to prepare. Sandwiches, one of the most popular snacks, originated in Europe. If they are eaten in place of a main meal, as in a packed lunch, they should contain the nutrients necessary to make up a balanced meal. Light dishes, snacks and sandwiches are particularly appropriate for picnics and parties.

Sweet Potato Sausages *Pamela Malebe*

Serves 4

Ingredients

2 medium-sized sweet potatoes
1 tsp salt
450 g sausagemeat or 8 sausages, skinned
1 tbsp cooking oil

Method

1 Grate the raw sweet potatoes.
2 Mix them with salt and sausagemeat and shape into flat round cakes.
3 Heat the oil and fry the cakes for about 5 minutes until golden brown on both sides. Serve hot.

Cassava Cheese Balls *Mary Makokha*

Makes 12 balls

Ingredients

2 medium cassavas
1 tbsp chopped onion
2 cups grated cheese
pinch of salt
1 egg, beaten
2 cups breadcrumbs
fat for deep-frying

Light dishes and snacks 15

Method

1. Peel the cassavas, remove central stalks and boil until tender.
2. Sauté the chopped onion. Add cassava, grated cheese and salt and mash until smooth.
3. Shape into balls, dip into the beaten egg and roll in the breadcrumbs.
4. Deep-fry until golden brown, drain on absorbent paper and serve hot.

Savoury Arrowroot Balls   *Margaret Mungai*

Makes 12 balls

Ingredients

2 medium arrowroots
1 level tsp salt
1 medium onion
50 g minced beef
$\frac{1}{4}$ tsp black pepper
3 eggs
3 cups breadcrumbs
oil for deep-frying

Method

1. Peel the arrowroots, wash them and put them to boil in salted water.
2. Chop the onion and fry lightly. Add the minced meat and cook until brown. Remove from the heat.
3. When the arrowroots are cooked, mash them with the meat, onions, pepper and one egg. Beat the other two eggs.
4. Add 1 tbsp breadcrumbs to the arrowroot mixture and mash to a consistency that will not stick to floured hands.
5. Form into balls and coat first with beaten egg, then with breadcrumbs.
6. Deep-fry until golden brown. They can be served hot or cold, as part of a meal or with tea. The balls can be made very small and served as appetizers.

Sweet Potato Balls   *Gertrude Ireri*

Serves 3

Ingredients

2 medium-sized sweet potatoes
2 tbsp flour
$\frac{1}{4}$ tsp salt
1 tsp white pepper
1 egg, beaten
oil for deep-frying

Method

1 Boil the sweet potatoes in their skins until tender. Drain, peel and mash.
2 Combine the sweet potatoes with the flour, salt, pepper and beaten egg. Mix well.
3 Form into balls about 4 cm in diameter.
4 Fry in deep fat until brown. Drain well and serve hot.

Sweet Potato Meat Balls
Damari Otieno

Serves 4

Ingredients

4 medium-sized sweet potatoes
1 medium onion
1 cup minced meat
pinch of salt
1 tsp ground cloves
1 tsp black pepper
1 clove garlic
1 tsp mixed spices
1 tbsp flour
2 tbsps water
1 egg, beaten
1 cup breadcrumbs
oil for deep-frying
lettuce and tomatoes to garnish

Method

1 Peel and chop the sweet potatoes. Put them into boiling water and boil for about 20 minutes.
2 Meanwhile chop and fry the onion in 1 tsp fat until golden brown.
3 Season the meat with the salt and spices. Add to the onion and fry for 10 minutes.
4 Make a paste with the flour and water and stir into the meat to bind it.
5 Mash the potatoes and shape into balls. Make a hole in the centre of each ball, fill it with the meat mixture, then close.
6 Coat first with the beaten egg, then with breadcrumbs and fry in deep fat for 5 minutes. Serve, garnished with tomatoes, on a bed of lettuce.

Light dishes and snacks 17

Brown Potato Balls
Betty Maina

Serves 4

Ingredients

½ kg potatoes
2 egg yolks
2 tbsps margarine
4 tsps chopped dhania
¼ tsp pepper
1 tsp salt
a little flour
2 whole eggs, beaten
1 cup breadcrumbs
parsley and tomatoes to garnish

Method

1. Boil and mash the potatoes.
2. Add the two egg yolks, margarine, chopped dhania, pepper and salt. Mix well.
3. Form into balls on a floured board and coat first with the beaten eggs, then with the breadcrumbs.
4. Fry in deep fat until golden brown. Drain well. Serve garnished with parsley and tomato slices.

Spiced Sweet Potatoes
Rose Osungu

Serves 4

Ingredients

4 sweet potatoes
1 tsp flour
2 tbsp sugar
1 tsp salt
1 tsp cinnamon
2 tbsps margarine
grated carrot and parsley to garnish

Method

1. Peel the sweet potatoes and cut each into four slices.
2. Put the potatoes in a heavy saucepan and cover with water.
3. Bring to the boil, cover tightly and simmer for 45 minutes or until soft but intact. Drain off the liquid and mix it with 1 cup of water and the flour, sugar, salt, cinnamon and margarine to make a sauce. Add more water if necessary. Boil the sauce for 3–5 minutes.

4 Pour the sauce over the sweet potatoes. Decorate with grated carrots and parsley and serve at once.

Sweet Potato Cakes
Jane Kibuga

Serves 4

Ingredients

1 sweet potato
$\frac{1}{2}$ tsp salt
$\frac{1}{4}$ tsp pepper
1 egg yolk
1 tsp margarine
2 tbsp white flour
1 tbsp fat

Method

1 Peel, wash, boil and mash the sweet potato.
2 Mix the potato with the salt, pepper, egg yolk and melted margarine, then beat in the flour to form a stiff dough.
3 Roll out to 1 cm thickness and cut into four pieces.
4 Fry on a greased hot plate or frying pan until golden brown.

Sweet Potato Burgers
Annie Maina

Serves 4

Ingredients

2 medium-sized sweet potatoes
1 medium onion, chopped
1 clove garlic
125 g minced meat
$\frac{1}{4}$ tsp salt
$\frac{1}{8}$ tsp black pepper
3 eggs
1 cup breadcrumbs
fat for deep-frying

Method

1 Boil the sweet potatoes until soft.
2 Sauté the onion and garlic, then add the meat, salt and pepper. Cook the mixture, stirring continuously, for about 5 minutes.

3 Peel the sweet potatoes and mash with the meat mixture, one egg and $\frac{1}{2}$ tbsp breadcrumbs. Beat the other two eggs.
4 Form into small balls, dip them in the beaten egg and coat with breadcrumbs. Deep-fry until golden brown. Serve hot or cold.

Sweet Potato Crisps

Elizabeth Kagwe

Serves 4

Ingredients

2 large sweet potatoes
oil for deep-frying
$\frac{1}{2}$ tsp salt

Method

1 Wash the sweet potatoes and boil for 10 minutes or until tender. Remove from the heat, peel and slice finely.
2 Deep-fry until golden brown (about 7 minutes).
3 Put the crisps on absorbent paper and sprinkle with salt to remove excess oil. Cover with paper for 2 minutes. Serve hot.

Potato Delight

Rahab Kamau

Serves 4

Ingredients

4 potatoes
2 cups coconut milk
pinch of salt
$\frac{1}{4}$ cup grated Cheddar cheese
1 tbsp margarine
$\frac{1}{2}$ cup chopped runner beans
$\frac{1}{2}$ medium carrot, diced

Method

1 Peel the potatoes and boil them in the coconut milk.
2 Add the salt, cheese and margarine and mash until smooth.
3 Pipe the potatoes round a flameproof plate and grill for about 2 minutes or until golden brown. Remove and garnish with steamed runner beans and carrots. Serve immediately.

Cassava Cakes
Jane Kibuga

Serves 4

Ingredients

1 small cassava
1 tbsp margarine
1½ tbsps flour
¼ tsp salt
1 tbsp fat
1 medium onion
1 tbsp grated cheese

Method

1. Boil and mash the cassava and mix it with the margarine, flour and salt.
2. Roll out on a floured board and cut into circles.
3. Heat the oil in a frying pan and fry the cakes until golden brown on both sides.
4. Cut the onion into rings, coat with grated cheese and place on top of each cake. Cook under the grill until the cheese melts. Serve hot.

Kaimati
Maria Onyango

Makes 16

Ingredients

½ tsp dry yeast
1 tbsp sugar
1 egg, beaten
1 cup water or milk
2 cups flour
1 tsp salt
oil for deep-frying

Method

1. Cream the yeast with a little sugar, the beaten egg and water or milk.
2. Sieve the flour and salt into a bowl.
3. Make a well in the centre of the dry ingredients and pour in the liquid. Mix together and beat until smooth. Cover with a damp cloth and leave to rise for 1½–2 hours.
4. Deep-fry the mixture in spoonfuls until golden brown. Roll in sugar. Serve hot or cold.

Light dishes and snacks 21

Matoke Sandwich
Rachel Burugu

Serves 2

Ingredients

4 green bananas (matoke)
1 cup flour
2 tsps salt
1 tsp pepper
1 large onion
4 tbsps fat
250 g minced beef
½ cup breadcrumbs
2 eggs

Method

1. Peel, boil and mash the bananas and mix with the flour, salt and pepper.
2. Put the mixture onto a floured board and roll out. Cut into circles and fry. Keep warm.
3. Chop the onion and fry the meat until brown.
4. Mix together the onion, breadcrumbs and eggs.
5. Add the mixture to the meat and cook for 30 minutes, stirring all the time.
6. Form into circles of the same size as the matoke cakes. Sandwich the meat between the matoke cakes. Serve hot with vegetables or salad as a snack or light lunch.

Stuffed Avocado Pears
Betty Maina

Serves 4

Ingredients

6 bacon rashers
2 ripe avocado pears
1 lemon
1 tbsp mayonnaise
1 tbsp grated cheddar cheese
1 small red pepper

Method

1. Trim the rind from the bacon, cut the rashers into very small pieces and fry them in their own fat.
2. Meanwhile, halve the avocado pears and remove the stones.
3. Squeeze the juice from the lemon and sprinkle it onto the halved avocados to prevent darkening.

4 When the bacon is ready, put it into a bowl, add mayonnaise and mix well. Put the stuffing into each half of the avocado pears, sprinkle the top with grated cheese and garnish with slices of red pepper.

Sausage Potatoes   *Dorcas Male*

Serves 4

Ingredients

4 medium potatoes
4 sausages
1 egg yolk
$\frac{1}{2}$ tsp salt
$\frac{1}{2}$ tsp pepper
1 tbsp margarine
$\frac{1}{4}$ cup milk
parsley to garnish

Method

1 Peel the potatoes and boil them until soft.
2 Prick the sausages. Simmer them in $\frac{1}{2}$ cup water for 20 minutes. Skin them and cut into halves.
3 Mash the potatoes with the egg yolk, pepper, margarine and milk.
4 Take a spoonful of the potato, lay a piece of sausage on it, cover with potato, leaving it quite rough. Continue until all the sausages and potato have been used up. Place on a greased baking tin and bake at 350 °F (180 °C, gas mark 4) for 20 minutes or until brown. Garnish with parsley and serve with tomato or other sauce.

Soup

Soup is a liquid food based on one or a combination of ingredients: meat, vegetables or pulses. It may be served as a starter or as a meal in itself, with bread, ugali or chapatis. Hot, thick soup is nourishing and warming at cooler times of the year.

Soups can be classified into four main types. Clear soup or consommé is flavoured, clear stock with a garnish. Broth contains small pieces of meat and/or vegetables, sometimes with rice, barley or small pieces of pasta. A thickened soup is one that has a thickening ingredient such as flour or egg yolk added to it. A purée contains ingredients that thicken the soup by themselves: the soup is passed through a sieve to make the texture smooth.

The liquid in soup is usually stock (which can also be used in sauces). Stock is made by simmering bones, fish or meat scraps or vegetables in water for at least 3 or 4 hours. (Fish stock should be simmered for a maximum of 45 minutes – cooking it for longer tends to make it taste bitter.) The stock should be cooled as quickly as possible and kept in a cool place, ideally in a refrigerator. The fat settles and solidifies on the top and should be removed before the stock is used. A more convenient alternative to making stock is to dissolve stock cubes in hot water.

A wide variety of soup can be made by choosing and experimenting with different combinations of ingredients and preparing them in different ways. Herbs and spices may be used to add to the flavour.

Bone Soup *Elizabeth Karani*

Serves 4

Ingredients

a root from a medicinal tree (e.g. muteta)
6   cups water
$2\frac{1}{2}$ kg oxbones
2   bay leaves
$1\frac{1}{2}$ cups hot milk
salt to taste

Method

1 Boil the roots in 3 cups of water until they are the colour of black coffee or the usual colour of the root. Strain the liquid into another container.

2 Boil the bones and bay leaves in 3 cups of water for about 15 minutes, then cover and simmer for about an hour.
3 Strain the soup, pour in the hot milk and add salt to taste.
4 Add 1 cup of root liquid and whisk thoroughly. Serve hot.

Beef Broth
Pauline Mudek

Serves 2

Ingredients

500g beef bones
pinch of salt
½ cup milk
parsley to garnish

Method

1 Wash the bones and put them into a saucepan just large enough to hold them. Cover them with water.
2 Bring the water very slowly to boiling point, then simmer for 1 hour. Add salt.
3 Drain the soup from the bones and simmer for 5 minutes.
4 Whisk the soup for 5-7 minutes. Add milk, whisk again and simmer briefly. Garnish with parsley to serve.

Vegetable Soup
Laura Karanja

Serves 4

Ingredients

3 onions
3 potatoes
3 carrots
5 tbsps margarine
8 cups water
1 tsp mixed herbs
¼ cup parsley
2 tsps salt

Method

1 Peel and chop the onions, potatoes and carrots.
2 Melt 4 tbsps margarine in a saucepan, add the onions and fry until brown.
3 Add water, carrots, potatoes, mixed herbs, parsley and salt.

4 Bring to the boil, then lower the heat. Cover and simmer until the vegetables are tender (about 35 minutes).
5 Pass through a sieve, then add 1 tbsp margarine and heat for about 3 minutes. Serve hot.

Brown Vegetable Soup   *Ruth Asiachi*

Serves 3

Ingredients

1 large carrot
1 turnip
5 celery stalks
1 tbsp fat
1 tbsp flour
2 tomatoes
2 beef stock cubes dissolved in 3 cups water
$\frac{1}{4}$ tsp ground garlic
$\frac{1}{4}$ tsp cumin seeds
$\frac{1}{2}$ tsp mixed herbs
salt to taste
parsley to garnish

Method

1 Prepare the vegetables and cut them into large pieces.
2 Melt the fat. When it starts to smoke add the vegetables (except the tomatoes) and fry until light brown.
3 Lift them from the fat and add flour to it to make a brown roux.
4 Put back the vegetables and add the chopped tomatoes, stock and spices.
5 Cook until tender, then liquidise or pass through a sieve. Season to taste, garnish with parsley and serve hot.

Cream of Tomato Soup   *Joyce Kimaru*

Serves 4

Ingredients

1 tbsp butter
5 medium tomatoes
1 onion
2 bay leaves
4 fresh peppercorns
1 stock cube dissolved in 1 cup water

2 heaped tbsps flour
3 cups milk
salt to taste

Method

1. Heat the butter and fry the sliced tomatoes, chopped onion, bay leaves and peppercorns.
2. Add the stock and simmer until the vegetables are soft.
3. Pass through a sieve and return to the heat.
4. Dissolve the flour in the milk, then add to the strained vegetables, stirring constantly. Add salt to taste. Serve hot.

Sutek
Leah Kibiego

Serves 2

Ingredients

250 g bones (preferably from the hip joint)
1 carrot
1 spring onion
1 tbsp wimbi flour

Method

1. Boil the bones and carrot together in 4 cups of water for 15-20 minutes.
2. Add the spring onion and boil for 2-4 minutes.
3. Remove the bones and carrot and sieve the stock; it should give 2-3 cups.
4. Blend the wimbi flour with a little water until smooth.
5. Add the wimbi paste to the boiling stock, stirring continuously to prevent lumps forming; boil for 5-7 minutes. The soup should be thin in consistency. Serve hot.

Variation: other vegetables (e.g. potatoes) may be added at stage 1.

Eggs

Eggs are very useful in cooking because they can be eaten on their own, cooked in several different ways, and they are also an important ingredient of many other dishes, such as cakes, sauces and batters for making pancakes etc. Eggs are very nutritious, as they contain protein (in the yolk), easily digested fat, vitamins A, B_1, B_2 and D, iron, calcium and phosphorous. They should always be used fresh, as their nutritional value decreases with age. When eaten by themselves, they are most commonly boiled, scrambled, poached or fried.

Boiled eggs

Put the eggs into enough boiling water to cover them and bring the water back to the boil. Boil gently for 3–5 minutes (for a soft-boiled egg) or for 7–10 minutes (for a hard-boiled egg). Cool hard-boiled eggs immediately in cold water to prevent the inside darkening.

Scrambled eggs

Beat the eggs with 1 tbsp milk for each egg and add salt and pepper. Over a moderate heat, melt 1 tbsp butter or other fat for each egg. Pour the eggs into the pan and cook, stirring constantly, until they are set but still soft. Serve immediately on toast.

Poached eggs

Bring a shallow saucepan of salted water to the boil. Break the eggs and slide them gently into the water. Simmer until the white has set. Drain well and serve on toast.

Fried eggs

Allow about 2 tsps fat for each egg. Heat the fat and break the egg into a cup or saucer. Slide the egg into the hot fat, and fry gently, basting with the fat. Drain off any excess fat before serving.

Pancakes

Pauline Mudek

Serves 3

Ingredients

3 eggs
1½ cups plain flour
½ cup sugar
3 tbsps baking powder
pinch of salt
1 cup milk

Method

1 Beat the eggs thoroughly.
2 Sieve the dry ingredients into a bowl. Make a hole in the centre and drop in the beaten eggs.
3 Add just enough milk (about 4 tbsps) to incorporate all the flour. Stir until the mixture is quite smooth.
4 Add the remaining milk gradually, stirring constantly.
5 Beat well until the surface of the batter is covered with tiny air bubbles.
6 Put a small piece of cooking fat into the frying pan and heat until it smokes. Pour in enough batter to cover the bottom of the pan. Cook until it can be shaken free of the pan. Turn over and cook until light brown. Serve dredged with sugar.

Sweet Potato Pancakes

Eva Munene

Serves 4

Ingredients

1 cup flour
1 tsp baking powder
1 tbsp sugar
1 medium-sized sweet potato
juice of half a lemon
1 egg
2 tbsps milk or water
2 tbsps salad oil or dripping

Method

1 Sieve the flour, baking powder and sugar into a large mixing bowl.
2 Boil and mash the sweet potato and add to the flour. Mix thoroughly and make a well in the centre.
3 Add the beaten egg and mix into a batter, adding milk or water until the

batter pours smoothly from the back of a spoon. Add the lemon juice.
4 Fry the pancakes in a little salad oil until golden brown on both sides.
5 Roll the pancakes and serve hot with jam or honey.
Variation: add a pinch of cinnamon to the flour.

Carrot Pancakes
Helen Wandera

Serves 4

Ingredients

1 cup flour
1 tsp baking powder
$\frac{1}{4}$ small carrot, grated
1 tbsp sugar
pinch of salt
2 eggs, beaten
6 tsps oil
7 tbsps milk

Method

1 Sieve the flour, salt and baking powder together.
2 Add the grated carrot and sugar and mix thoroughly. Make a well in the middle of the mixture and add the milk and the beaten eggs.
3 Fold in the flour and mix to a smooth dropping consistency.
4 Heat the oil in a pan. Drop the batter into it using a dessertspoon. Cook the pancakes on both sides until golden brown.
5 Remove the pancakes from the pan and put them on greaseproof paper before serving.
Variation: use sour milk, $\frac{1}{4}$ tsp bicarbonate of soda and $\frac{1}{4}$ tsp baking powder. This produces very soft and spongy pancakes because of the acid in the sour milk.

Sour Milk Pancakes
Leah Marangu

Serves 6

Ingredients

2 cups flour
$\frac{1}{4}$ cup sugar
1 tsp bicarbonate of soda
1 tsp salt
$2\frac{1}{4}$ cups sour milk (maziwa lala)
2 eggs, beaten
$\frac{1}{4}$ cup melted fat or salad oil

Method

1 Sift the dry ingredients into a bowl. Make a well in the middle. Add maziwa lala, beaten eggs and oil and stir gently until all ingredients are blended: do not overbeat.
2 Heat a little fat in a skillet until a little water dropped into it sizzles.
3 Drop batter in large spoonfuls into the heated skillet and fry until bubbles form on top of the pancakes before you turn them.
4 Turn and fry the other side until golden brown and dry in the middle when pricked with a fork. Serve with syrup or sprinkled with sugar.

Arrowroot Pancakes *Fracia Kibugu*

Serves 3

Ingredients

1 medium arrowroot
pinch of salt
1 cup flour
2 eggs, beaten
1 tbsp sugar
$\frac{1}{2}$ cup corn oil

Method

1 Wash the arrowroot and boil it in salted water.
2 Beat together the flour, eggs, sugar and salt to a thin consistency.
3 Slice the cooked arrowroots. Cover the slices with batter and fry until golden brown. Serve hot.

Green Maize Omelette *Cecilia Nderitu*

Serves 4

Ingredients

2 cups cooked green maize
$\frac{1}{2}$ tsp salt
$\frac{1}{8}$ tsp pepper
2 tbsps milk
2 eggs, beaten
1 tbsp margarine

Method

1 Pound the maize or blend in a liquidiser or mincer until fine. Mix with salt, pepper, milk and eggs.
2 Fry in margarine until golden brown on both sides (about 3 minutes). Serve hot.

Mixed Vegetable Omelette
Eva Gatere

Serves 3

Ingredients

1 small onion
3 small cooked potatoes
1 red pepper
1 large tomato
1 clove garlic
2 tbsps oil
3 tbsps cooked green peas
3 eggs
seasoning
pinch of mixed herbs

Method

1 Finely chop the onion, dice the potatoes, pepper and skinned tomato (discard seeds) and crush the garlic.
2 Heat 1 tbsp oil and fry the onion and garlic until soft.
3 Add the remaining oil, then stir in the rest of the vegetables and heat thoroughly for 8–10 minutes.
4 Beat the eggs lightly, add seasoning and herbs and pour over the vegetables.
5 Stir lightly and leave on the heat until the eggs are just set. Serve hot.

Scotch Eggs
Joyce Onyango

Makes 2

Ingredients

3 eggs
1 tbsp plain flour
200 g sausagemeat or 4 sausages, skinned
4 tbsps breadcrumbs
fat for deep-frying

Method

1. Hard boil two eggs and allow them to cool. Peel them and dust lightly with plain flour. Beat the other egg.
2. Divide the sausagemeat into two. Shape the meat into flat cakes.
3. Wrap each egg evenly in the sausagemeat.
4. Coat them in the beaten egg and then roll each in the breadcrumbs.
5. Heat the fat to about 340 °F, or until an inch cube of bread turns golden brown in one minute.
6. Fry the eggs for 7–10 minutes. Halve them with a sharp knife and serve cold with a salad or hot with a juicy vegetable such as tomatoes.

Variation: before coating the eggs with sausagemeat, halve them and scoop out the yolks. Mix the yolks with $\frac{1}{8}$ tsp nutmeg, $\frac{1}{8}$ tsp mixed spice, $\frac{1}{8}$ tsp black pepper, $\frac{1}{8}$ tsp mustard powder and $\frac{1}{4}$ tsp butter. Scoop the seasoned yolks into the egg whites and coat with sausagemeat.

One-pot dishes

Most main dishes are one-pot dishes. Making a main dish in a single pot is common in many Kenyan homes with only one fire to do the cooking on. A one-pot dish should contain all the nutrients needed in one meal: body-building, energy-giving and protective elements. A dish made with potatoes, meat, tomatoes and onions cooked together is an example of such a meal. Some dishes that are poor in nutrients should be eaten with green vegetables.

These dishes can be based on any of the staple foods. For instance, dishes based on potatoes or green bananas are best combined with animal protein foods. Those based on cereals such as maize or wheat will be complete when combined with pulses, nuts, vegetables or meat.

Yam Potage *Grace Kogi*

Serves 4

Ingredients

1 medium yam
$\frac{1}{2}$ kg beef
2 tbsps chopped fresh dhania
1 medium onion
2 tbsps cooking fat
3 medium tomatoes
3 carrots
2 cups water or stock
salt and pepper to taste

Method

1. Peel and cut the yam into 1 cm cubes and cover with salted water.
2. Clean and chop the meat into pieces the same size as the yam.
3. Finely chop the dhania and onion and fry with the meat in a tightly covered saucepan.
4. Add the skinned and chopped tomatoes. Stir the contents and cook for 10 minutes.
5. Drain the yam, add to the meat and cook for a further 5 minutes.
6. Scrape and chop the carrots and add them to the other ingredients.
7. Add the water or stock; stir and cook for 10 minutes. Serve hot.

Stewed Bananas

Fracia Kibugu

Serves 2

Ingredients

8 green bananas
1 tbsp margarine or ghee
100 g minced meat
1 onion
1 clove garlic
1 green pepper
3 leaves spinach, chopped
2 tsps salt
1 large tomato, chopped

Method

1. Peel the bananas and soak in salted water.
2. Fry the minced meat and chopped onion, garlic and green pepper on a low heat.
3. Wash and chop the spinach.
4. Wash the bananas and add to the meat mixture. Cover the pan.
5. After about 20 minutes, add the spinach, salt and tomato. Cook for a further 5 minutes.

Firinda

Dolrosa Ouma

(Sweet Potatoes and Beans)

Serves 2

Ingredients

1 cup beans
2 sweet potatoes
1 onion
2 tomatoes
2 tbsps fat
1½ tsps salt
1 cup coconut milk

Method

1. Soak the beans for 12 hours. Remove their skins and boil until soft.
2. Wash, peel and slice the sweet potatoes into pieces about 1 cm thick. Boil until tender.
3. Strain the water from the beans and put it aside. Mash the beans or pass them through a sieve.

4 Finely chop the onion and tomatoes. Melt the fat, add the onions and fry until they start browning. Add the tomatoes and salt.
5 Mix the water strained from the beans with the mashed beans and add this to the frying onions and tomatoes. Leave to simmer for 3 minutes, add the coconut milk and simmer for 10 minutes.
6 Drain the water from the potatoes and pour the bean sauce over them. Serve hot.

Mucui
Charity Njiru

Serves 4

Ingredients

1 medium yam
1 medium arrowroot
4 medium potatoes
2 medium green bananas
2 medium carrots
250 g beef
1 medium onion
1 tbsp kimbo
1 cup green peas
$\frac{1}{4}$ tsp mixed spice
pinch of salt
$\frac{1}{4}$ tsp white pepper

Method

1 Peel, wash and chop the yam, arrowroot, potatoes, green bananas and carrots.
2 Cut the meat into small pieces and slice the onion.
3 Fry the onion until golden brown. Add the meat and cook until tender (about 60 minutes).
4 Add the peas and carrots and cook for 10 minutes.
5 Add the yam, arrowroot, bananas and potatoes. Cook for 5 minutes.
6 Add the spices, salt and pepper and cook for 5 minutes.
7 Add water to cover and leave to cook until tender (about 30 minutes). Serve hot.

Rumonde
Alice Kirigia
(Marigu ja Mwana)

Serves 2

Ingredients

2   medium potatoes
2   green bananas

125 g beef
1 medium onion
1 tbsp fat
1 tomato
1½ cups water

Method

1. Peel the potatoes and bananas and cut into small pieces.
2. Cut the meat into small pieces and chop the onion.
3. Fry the onion in the fat until light brown. Add the meat and cook for 5–10 minutes.
4. Add the blanched and chopped tomato, bananas, potatoes and enough water to cover the contents. Add salt and pepper.
5. Boil for 5–10 minutes, then simmer until tender. (If the water dries up add a little more.) Drain and mash the mixture.

Variation: 1 tsp chopped dhania may be added at stage 4.

Beef and Vegetable Special *Mary Ndegwa*

Serves 2

Ingredients

1 cup shelled green peas
250 g steak
1 medium onion
1 bunch dhania
1 tomato
1 tbsp fat
1 tsp margarine
½ tsp curry powder
4 medium carrots
3 medium potatoes
1 tsp salt
1 clove garlic

Method

1. Wash and boil the peas.
2. Cut the meat into small pieces, chop the onion and dhania, skin and chop the tomato.
3. Fry the onion and dhania together in the fat and margarine.
4. Add the meat, tomato and curry powder; cover and cook, stirring occasionally, for about 40 minutes.
5. Wash, scrape and dice the carrots and add them to the meat. Peel and dice the potatoes, keeping them in water until needed.
6. When the meat is cooked, add the potatoes, peas and salt and enough water to cover the contents.

7 Crush the garlic and add to the mixture. Cook until the potatoes are soft. Serve hot.

Ngunja Gutu
Eva Gatere

Serves 4

Ingredients

½ tbsp cooking fat
1 medium onion
¼ tsp mixed spice
½ tsp curry powder
2 small potatoes
½ bunch pumpkin leaves
1 tsp salt
2 cups water
1 cup cooked pigeon peas
1 tomato
½ cup maize flour

Method

1 Heat the fat, add the finely chopped onion, mixed spice and curry powder and cook until the onion is soft.
2 Add the chopped potatoes and cook for 10 minutes, adding a little of the water if necessary. Add the greens and salt and simmer for 15 minutes.
3 Add the cooked pigeon peas and chopped tomato and heat thoroughly.
4 Mash the mixture, adding a little flour at a time. Cover and cook for 15 minutes, stirring frequently. Serve with meat stew.

Ngunja Gutu with Meat
Cecilia Nderitu

Serves 2

Ingredients

1 small onion
1 tbsp fat
250 g minced meat
4 cups water
½ bunch pumpkin leaves
1 large carrot, grated
1 tsp mixed spice
salt to taste
1 cup maizemeal

Method

1. Chop the onion finely and fry until brown. Add the meat, cover and cook for 10 minutes. Add 2 cups water and boil for 40 minutes.
2. Wash and chop the greens, finely grate the carrot. Add to the meat with 1 cup water and cook until soft. Add the spice and salt.
3. Add 1 cup of water and bring to the boil. Add the maizemeal and cook on a low heat for 20 minutes. Serve hot.

Miraaru Hotpot

Agnes Karumba

Serves 4

Ingredients

300 g stewing steak
8 medium bananas (miraaru)
1 medium onion
2 tbsps fat
2 medium tomatoes
1 cup shelled peas
pinch of black pepper
2 tsps salt
$1\frac{1}{2}$ cups water

Method

1. Remove excess fat from the meat and cut into cubes. Peel the bananas and cover them with water.
2. Chop the onion finely and fry gently in the fat. Add the meat and fry until brown.
3. Blanch and chop the tomatoes and add to the meat with the peas, bananas, water and pepper.
4. Bring to the boil, then lower the heat and simmer until the meat and bananas are tender (about 45 minutes).
5. Season well and serve hot.

Note: if miraaru are not available, matoke can be used. These are especially suitable since they are soft, and turn yellow when cooked. Miraaru are firmer and do not cook as quickly as matoke.

Kitoweo

Rosina Mutia

Serves 4

Ingredients

1 onion
1 small green pepper

1 tbsp shortening
1 tsp curry powder
1 tbsp beef roiko mchuzi mix
pepper to taste
3 tomatoes
250 g meat
3 carrots
5 green bananas
6 potatoes

Method

1 Chop the onion and pepper and fry them in the shortening until soft. Add the curry powder, pepper, roiko and chopped tomatoes, meat and carrots. Cook for 40 minutes.
2 Add the peeled and chopped bananas and potatoes and cook for 20 minutes. Serve hot.

Matumbo Mash
Elizabeth Murugu

Serves 4

Ingredients

250 g matumbo
8 medium potatoes
8 green bananas (matoke)
1 onion
1 tbsp fat
1 tsp curry powder
salt to taste
parsley or tomato to garnish

Method

1 Wash the matumbo and cut into small pieces. Pressure-cook for 15 minutes or until soft.
2 Peel the potatoes and bananas and chop the onion. Fry the onion until light brown.
3 Add curry powder, matumbo, potatoes and bananas and enough water to cover. Cook until the potatoes and bananas are soft.
4 Mash the mixture to a smooth consistency. Serve hot, garnished with parsley or tomato.

Mashed Bananas
(Lumonde)

Mary Mbae

Serves 4

Ingredients

4 green bananas (kiganda)
7 medium potatoes
1 medium onion
1½ tbsps fat
1 kg beef steak
1 cup shelled green peas
2 medium tomatoes
salt to taste

Method

1. Peel and wash the bananas and potatoes.
2. Fry the chopped onion on a low heat until golden brown.
3. Cut meat and tomatoes into small pieces and add to the onions. Lower the heat. Cover the saucepan and cook for 10 minutes.
4. Add the bananas, potatoes and peas to the meat, with enough water to cover them. Add salt.
5. Cover the saucepan and cook for 20–30 minutes.
6. Mash until soft. Serve hot.

Mchanyato

Marie Rebman

Serves 2

Ingredients

6 green cooking bananas
2 tomatoes
1 onion
½ tbsp fat
125 g meat
3⅓ cups coconut milk
½ tsp white pepper

Method

1. Peel the bananas, blanch and chop the tomatoes and chop the onion.
2. Fry the onion until golden brown. Add the chopped meat, tomatoes and bananas and cook for one hour in half the coconut milk.
3. Add the remaining coconut milk and simmer until tender.
4. Add seasoning and mash.

Variation: omit the coconut milk and add ¼ cup milk and 1½ cups cooked green peas at stage 2 and 2 tbsps margarine at stage 4.

Spaghetti and Cheese

Loxana Mwangi

Serves 1

Ingredients

100 g spaghetti
¼ tsp salt
1 egg
½ tbsp chopped onion
2 small tomatoes
1 clove garlic
1 tsp margarine
¼ tsp white pepper
1 tbsp grated cheese
parsley to garnish

Method

1. Boil the spaghetti in salted water. Hard boil the egg.
2. Sauté the chopped onion, tomatoes and garlic until they form a thick sauce.
3. Add the cooked spaghetti and pepper to the sauce. Simmer for 5 minutes, sprinkle with grated cheese and serve hot, garnished with parsley and slices of egg. May be accompanied with salad.

Cheese Eggplant Casserole

Gacheri Mburugu

Serves 2

Ingredients

1 eggplant
2 tbsps margarine or butter
½ tsp salt
1 tbsp flour
¼ tsp pepper
1 small onion, finely chopped
½ tsp oregano
2 large tomatoes, sliced
¼ cup grated Cheddar cheese

Method

1. Peel and slice the eggplant and boil in salted water for 10 minutes. Drain.
2. Mix the melted fat, salt, flour, pepper, onion and oregano to a paste.
3. Coat the eggplant slices with the paste and put them in a greased casserole.
4. Cover with tomato slices and sprinkle with cheese.
5. Bake in a moderately hot oven (190 °C, 375 °F, gas mark 5) for 25 minutes. Serve hot.

Stews

Most stews are made by simmering the ingredients for a long period of time. They can be cooked in the oven (casseroles) or over direct heat. Stews are usually eaten with ugali, potatoes, green bananas or yams, and curries (often cooked in a similar way to stews) with chapatis and/or rice. The nutritional value of a stew will depend on its ingredients.

Bean Stew *Margaret Maina*

Serves 2

Ingredients

1 cup beans
$\frac{1}{2}$ cup shelled green peas
2 small carrots
1 small onion
1 tbsp fat
$\frac{1}{4}$ tsp black pepper
$\frac{3}{4}$ tsp salt
1 chicken stock cube

Method

1 Soak the beans overnight.
2 Boil the beans in 4 cups water until almost soft. Add the peas and boil for 5 minutes or until soft.
3 Prepare and chop the carrots and onion. Fry for 2 minutes. Add the pepper, salt, beans, peas and stock made with the stock cube and 2 cups of hot water. Boil until the beans and peas are soft. Serve hot with rice or ugali.

Meat Stew *Mary Atalitsa*

Serves 4

Ingredients

1 tbsp fat
1 large onion, chopped

$\frac{3}{4}$ tsp salt
$\frac{1}{2}$ tsp black pepper
300 g meat, chopped
2 tomatoes, chopped
1 large green pepper, chopped
2 cups water

Method

1. Heat the fat in a saucepan and fry the onion until soft. Add salt, black pepper and tomatoes and cook until soft.
2. Add the meat, cover and cook for 5 minutes. Add water and cover tightly.
3. Cook until the meat is tender (about 30 minutes). Add the green pepper and cook for 10 minutes.
4. Simmer for 5 minutes and serve hot.

Brown Meat Stew

Elizabeth Murugu

Serves 2

Ingredients

1 onion, chopped
1 tbsp fat
250 g stewing steak, chopped
1 large tomato, peeled and chopped
1 bunch dhania, chopped
1 green pepper, chopped
1 chilli pepper, chopped
2 cups water or stock
salt to taste

Method

1. Fry the chopped onion in fat until soft. Add the meat and fry until brown.
2. Add the tomato, dhania, chopped green pepper and chilli pepper. Cook until the mixture thickens.
3. Add water or stock and simmer for 90 minutes. Add salt to taste. Serve with ugali or rice.

Mixed Meat Stew

Mary Kamami

Serves 3

Ingredients

$1\frac{1}{2}$ cups peas
4 potatoes

1 tbsp fat
1 onion, chopped
250 g tender meat, chopped
1 green pepper, chopped
2 tomatoes, chopped
½ tsp salt
½ tsp curry powder
2 carrots, chopped

Method

1 Boil the peas. Peel the potatoes and leave them whole.
2 Fry the onion slowly, add the meat and cook on a low heat for 5 minutes. Add the green pepper, tomatoes, salt and curry powder and cook for a further 4–5 minutes.
3 Add the peas and enough water to cover. Boil for 5 minutes.
4 Add the potatoes and carrots and cook until soft. Serve hot.

Spicy Meat Stew *Anne Maina*

Serves 4

Ingredients

360 g beef steak
1 medium onion
1 medium potato
1 medium carrot
1 medium tomato
1 clove garlic
1 tbsp cooking fat
¼ tsp black pepper
1 tsp curry powder
¼ tsp mixed spice
1 tsp salt
3 cups water
parsley to garnish

Method

1 Chop the meat into small pieces.
2 Peel and dice the vegetables.
3 Sauté the chopped onion and garlic until light brown.
4 Add the meat, stir and fry until brown. Add all the seasonings and the vegetables and cook for about 5 minutes.
5 Add the water, let it boil briefly and simmer for about 60 minutes.
6 Serve hot, garnished with parsley.

Mutton Stew

Dinah Makokha

Serves 2

Ingredients

2 onions, sliced
2 tbsps fat
250 g mutton, cubed
1 tbsp tomato purée
2 medium potatoes, chopped
1 carrot, chopped
1 tbsp roiko mchuzi mix
3 cups water
1 tsp salt

Method

1. Fry the onions in hot fat until golden brown.
2. Add the mutton and fry for 15 minutes. Add the tomato purée, potatoes, carrot and mchuzi mix. Add the water and cook for 45 minutes.
3. Add salt and cook for a further 5-10 minutes. Serve hot.

Dried or Smoked Beef Stew

Estabel Mulimba

Serves 3

Ingredients

300 g dried or smoked beef
1 small onion, chopped
1 tbsp ghee
pinch of salt
pinch of bicarbonate of soda
1 tomato, chopped
1 small carrot, diced
1 bunch dhania
1 green pepper (optional)
pinch of white pepper
$\frac{1}{2}$ tsp cumin seeds
1 tsp roiko mchuzi mix (beef flavour)
$\frac{3}{4}$ tsp curry powder
$\frac{1}{2}$ cup water
parsley to garnish

Method

1. Chop the beef into small pieces and boil until soft.
2. Fry the onion in ghee until soft.

3 Add the beef and fry until brown, seasoning it with salt and a little bicarbonate of soda.
4 Add tomato, carrot, dhania and green pepper and fry, turning the mixture, for 2 minutes.
5 Add pepper, cumin seeds, roiko, curry powder and salt. Mix well.
6 Add ½ cup water and simmer for about 40 minutes. Garnish with parsley and serve hot with matoke or ugali.

Minced Beef Stew
Eva Gatere

Serves 2

Ingredients

1 tsp cooking fat
1 onion
1 clove garlic
¼ tsp mixed spice
¼ tsp curry powder
¼ tsp black pepper
dhania to taste
250 g minced beef
salt to taste
1 tbsp tomato purée
1 cup beef stock

Method

1 Heat the fat in a saucepan.
2 Add the finely chopped onion and garlic and cook until soft.
3 Add the spices, dhania and meat and simmer for about 5 minutes.
4 Add salt and tomato purée and simmer for 1 minute. Add 1 cup beef stock, bring to the boil and simmer for 10 minutes. Serve hot with spaghetti or rice.

Beef Stew with Roast Bananas
Dorcas Male

Serves 4

Ingredients

500 g beef steak
2 onions
2 green peppers
1 bunch dhania
4 ripe tomatoes
1 tbsp kimbo
1 tsp salt
4 cups water
8 green bananas

Method

1. Cut the meat into large cubes and chop the vegetables.
2. Heat the fat in a saucepan and fry the onions gently until transparent. Add the meat, stir well and add the salt. Let the meat brown.
3. Add the water and chopped vegetables, cover with a lid and bake at 350 °F (180 °C, gas mark 4) for about 1 hour.
4. Meanwhile, peel the bananas and roast them on a wire rack over red-hot charcoal. Serve with the stew.

Meat and Cabbage Stew   *Zipporah Kabugu*

Serves 2

Ingredients

250 g meat
2 medium potatoes
2 medium carrots
1 small cabbage
1 small onion
1 tbsp fat
1 cup shelled peas
1 tsp salt
$\frac{1}{4}$ tsp pepper
2 small tomatoes

Method

1. Wash the meat, dry it and chop into small pieces.
2. Dice the potatoes and carrots and slice the cabbage and onion.
3. Heat the fat in a saucepan, add the onion and fry until golden brown.
4. Add the meat, potatoes, carrots, cabbage, peas, salt and pepper. Blanch and chop the tomatoes and add to the stew.
5. Allow to cook for 3 minutes. Add a little water, bring to the boil and reduce the heat. Allow to simmer until the meat is tender. Serve hot.

Arrowroot Stew with Meat or Fish   *Josephine Mwatha*

Serves 2

Ingredients

250 g meat or fish
1 onion
$\frac{1}{3}$ cup fat
1 arrowroot

2 potatoes
1 large carrot
⅓ cup shelled peas
1 tomato
2 tsps salt
1 tsp white pepper

Method

1. Dice the meat (or fish) and onion and fry until the meat is brown.
2. Peel and dice the arrowroot, potatoes and carrot. Add them to the meat, with the peas. (If the peas are hard, parboil them first.)
3. Add the blanched and skinned tomato, salt and pepper. Add enough water to cover, and simmer for 60 minutes. If the water dries up, add a little more. Serve hot.

Note: any fish or meat can be used for this dish.

Sukumawiki Beef Stew

Mary Mulaku

Serves 2-3

Ingredients

1 onion, chopped
1 tbsp fat
300 g stewing steak
1 tomato, chopped
1 carrot, grated
1 bunch dhania
pinch of salt
pinch of black pepper
½ tsp curry powder
1½ bunches sukamawiki
1 cup water

Method

1. Sauté the onion in fat until brown.
2. Add the chopped steak and fry until brown.
3. Add the tomato, grated carrot, dhania, salt, pepper and curry powder. Stir.
4. Add the sukumawiki and mix it well with the beef.
5. Add the water and simmer for 15 minutes. The sukumawiki should remain green in colour. Serve with ugali.

Goulash

Leah Marangu

Serves 5

Ingredients

5 cups macaroni
500 g ground beef
1 large onion, chopped
2 tbsps margarine
½ cup tomato purée
¼ cup tomato sauce
1 tsp salt
pinch of freshly ground pepper

Method

1. Cook the macaroni, drain and keep it warm.
2. Brown the meat and onion in the margarine.
3. Add tomato purée, tomato sauce and seasoning. Mix thoroughly and simmer for 5–10 minutes.
4. Add macaroni, mix and heat thoroughly. If it is too dry, add a little water.
5. Serve with a crisp tossed salad.

Variation: spaghetti may be used instead of macaroni.

Beef and Bean Stew

Pauline Githiru

Serves 2

Ingredients

1 cup beans
250 g beef
1 onion
2 tbsps fat
1 carrot (optional)
1 tbsp flour
3 cups water

Method

1. Soak the beans overnight.
2. Cut the meat into small pieces, removing any undesirable sections. Skin and slice the onion.
3. Heat the fat and fry the onion and meat until they begin to brown. Scrape and chop the carrot.
4. Remove the onion and meat from the pan. Add the flour to the fat and cook over a gentle heat, stirring continuously, until brown.

5 Pour in the water and bring to the boil, stirring continuously. Add salt, onion, meat, carrots and beans.
6 Reduce the heat, cover with a tight lid and simmer the stew gently until the meat is tender (about 60 minutes). Serve hot with rice or ugali.

Arrowroot Stew *Margaret Muri*

Serves 2

Ingredients

1 medium arrowroot
1 medium potato
1 small onion
2 tbsps fat
3 cups water

Method

1 Chop the arrowroot and potato into small cubes. Keep them separate.
2 Chop the onion and fry in a saucepan until golden brown.
3 Add the potato and fry for 10 minutes.
4 Add the arrowroot and water and simmer for 10–15 minutes. Cook until the mixture thickens. Serve hot with meat and vegetables.

Green Banana Stew *Margaret Mungai*

Serves 2

Ingredients

6 green bananas
250 g stewing steak
1 medium onion
1 small green chilli
1 tomato
1 clove garlic
1 bunch dhania
1 tsp salt
$3\frac{1}{3}$ cups water or stock

Method

1 Peel and dice the bananas and cover with cold water.
2 Cut the beef into small pieces. Chop the onion, chilli and blanched tomato and crush the garlic. Remove stalks from the dhania and chop the leaves.
3 Sauté the onion and garlic until soft but not brown. Add the tomato, dhania and meat.
4 Keep stirring until the meat browns, then add the bananas and salt and stir for 3 minutes. Add the liquid, stir once and cover.
5 Cook for 45 minutes, stirring twice during cooking. Serve sizzling hot.

Yam Stew   *Eva Munene*

Serves 4

Ingredients

1 tbsp cooking fat
1 small onion, finely chopped
250 g beef, cubed
1 medium yam, diced
1 medium potato, diced
1 small carrot, grated
1 medium tomato, chopped
6 cups water
1 tsp curry powder
1 beef stock cube
$\frac{1}{2}$ cup shelled green peas
1 tsp salt
1 bunch dhania, chopped

Method

1 Melt the fat in a saucepan and add the chopped onion. Fry until light brown.
2 Add the meat, yam, potato, carrot and tomato and cook, stirring continuously, for 5 minutes over a low heat.
3 Add the water, curry powder, beef stock cube, green peas and salt. Cover the pan and leave to simmer for 30 minutes.
4 Add the dhania 5 minutes before the end of cooking time. Serve hot with rice, chapatis or on its own.

Pumpkin Stew
Margaret Mungai

Serves 2

Ingredients

4 sticks celery
1 onion
2 cups diced pumpkin
1 tbsp cooking oil
1 cup chopped stewing steak
3 cups water or stock
1 tsp salt
$\frac{1}{8}$ tsp black pepper

Method

1. Wash and chop the celery and onion. Heat the oil and sauté the celery and pumpkin for 3 minutes. Put them into a dish and cover.
2. Using the same oil, sauté the onion until tender but not brown.
3. Add the meat and cook until brown. Add the liquid, salt and pepper and cook for 15 minutes.
4. Add the pumpkin and celery and cook for a further 15 minutes. Serve hot.

Stewed Mushrooms
Dolrosa Ouma

Serves 3

Ingredients

$\frac{3}{4}$ cup dried mushrooms
1 medium onion
3 small tomatoes
2 tbsps ghee
$1\frac{1}{2}$ tsps salt
$1\frac{1}{2}$ cups fresh milk

Method

1. Wash the mushrooms and chop the onion and tomatoes.
2. Melt $1\frac{1}{2}$ tbsps of the ghee in a saucepan, add the onion and fry until golden brown. Add the tomatoes and salt and cook for 2 minutes.
3. Add the mushrooms and water. Cook for 10 minutes, then add the milk. Simmer for 20 minutes, then add the remaining $\frac{1}{2}$ tbsp ghee and simmer for 2 minutes. Serve hot with ugali, rice, cooked bananas or sweet potatoes.

Note: this dish is very popular among Western Kenyans.

Curried Meat and Vegetable Stew

Josephine Kavilu

Serves 2

Ingredients

1	tbsp fat
1	medium onion, chopped
1	tsp curry powder
200 g	stewing steak, cubed
1	cup shelled green peas
3	tomatoes, skinned and chopped
2-3	cups hot water
4	medium potatoes, diced
2	medium carrots, sliced
1½	tsps roiko mchuzi mix
1	sweet pepper, chopped

salt to taste

Method

1. Heat the fat in a saucepan and fry the onion until it is tender and starting to turn pale brown.
2. Add the curry powder and fry together until onions are golden brown. Add the meat and stir.
3. Simmer the meat, stirring occasionally, for about 20-25 minutes, adding a little water if it starts to go dry.
4. Add the peas and bring to the boil.
5. Add skinned and chopped tomatoes. Simmer for 5 minutes.
6. Add 1 cup of hot water and simmer for 25 minutes, stirring occasionally.
7. Add the potatoes and carrots. Stir and bring to the boil.
8. Sprinkle mchuzi mix into the stew and stir. Bring to the boil, then simmer for 5 minutes.
9. Add the remaining hot water and simmer until the potatoes are nearly cooked (15-20 minutes).
10. Add the chopped sweet pepper and salt to taste.
11. Simmer the stew until all the vegetables are tender and the pepper is a rich green colour. Serve hot.

Meat

Traditionally, people's meat-eating habits were strongly influenced by their customs, religion and taboos. Meat is a valuable source of good quality, easily digested proteins. It contains good amounts of the B vitamins, especially niacin and riboflavin, and is rich in iron and potassium. Its fat content ranges from 10% to 22%, so fattier meat can be a useful source of energy. However, excess fat should be trimmed off before cooking meat. Meat from wild game has similar nutritional value to that from domestic animals, but often has a stronger flavour.

Meat has to be cooked to make it tender, palatable and digestible. It can be prepared in many ways: roasting, stewing, boiling, frying and grilling. Marinating meat or rubbing it with pawpaw leaves before cooking helps to tenderise it.

Roasting is suitable for thick, tender cuts of meat. The meat is cooked briefly at a high temperature to seal in the juices, then cooked at a lower heat. It can be cooked in a covered container or left uncovered and basted from time to time. Stewing helps to tenderise tough meat: it is cooked slowly in liquid in a container with a tight-fitting lid. Boiling is the method most often used to cook salted meats. The meat is put into cold water, heated and allowed to simmer until cooked. Frying and grilling are suitable only for thin, tender pieces of meat. Meat to be fried should be put into hot fat, browned quickly on each side, then cooked.

Nyama Choma
(Barbecued Meat)

Rosina Mutia

Serves 2

Ingredients

250 g mutton, beef or pork
$\frac{1}{4}$ tsp black pepper
$\frac{1}{2}$ tbsp shortening
$\frac{1}{4}$ tsp salt

Method

1. Cut the meat into thin slices and sprinkle with black pepper.
2. Cook on both sides under a low grill, basting with fat to prevent it drying.
3. Sprinkle with salt and serve sizzling hot.

Marinated Nyama Choma

Joyce Meme

Marinade

- 1½ cups cooking oil
- ½ tsp salt
- ½ tsp black pepper
- ½ cup chopped dhania
- 1 diced onion
- 2 lemons, chopped

Method

1. Mix all the ingredients.
2. Put the meat into the dish and coat it well with the mixture.
3. Cover the dish and leave for twelve hours, turning the meat once or twice.
4. Remove meat and barbecue as normal, basting with any remaining marinade.

Note: marinating the meat helps to tenderise it.

Goat Ribs
(Faronte)

Florine Mbithi

Serves 4

Ingredients

- 500 g goat ribs (or beef, pork or lamb)
- 1 medium onion, chopped
- 1 tsp green ginger
- 1 hot pepper, chopped
- ¾ tsp masala
- salt to taste
- 1 tbsp tomato purée
- 3 tomatoes, chopped
- 1 bunch dhania, chopped
- ¾ tsp crushed garlic
- ¾ tsp paprika
- 1 tsp mixed herbs

Method

1. Cut the ribs into pieces.
2. Mix together all the other ingredients with a little water.
3. Spread the mixture over the pieces of meat so that they are well coated. Leave them to marinate overnight.
4. Grill the ribs over a hot charcoal grill. Serve hot.

Note: this is a good party dish.

Eva's Meatballs
Eva Munene

Serves 4

Ingredients

For the meatballs:
2 tbsps fat
1 tbsp chopped onion
½ cup chopped green pepper
½ cup chopped celery
1 tsp chopped dhania
1 tsp chopped parsley
250 g minced meat
pinch of salt
1 tbsp flour
1 egg, beaten
1 cup salad oil

For the sauce:
½ tbsp fat
1 small onion
1 tomato
1 tsp turmeric
salt and pepper
1 green chilli
½ pint water

Method

1 Melt the fat in a saucepan and fry the vegetables and herbs until tender. Allow to cool.
2 Put the minced meat in a bowl, add salt, flour, the beaten egg and the cooled vegetables.
3 Stir the mixture into a paste and shape into balls.
4 Fry the meatballs in salad oil for 25–30 minutes.

To make the sauce:
1 Melt the fat in a saucepan and fry the chopped onion and tomato until they are tender.
2 Add all the other ingredients. Boil for 10 minutes.
3 Put through a sieve and pour over the meatballs in a serving dish.

Sweet and Sour Meatballs
Leah Marangu

Serves 4

Ingredients

500 g ground beef
2 cups soft breadcrumbs
2 eggs, lightly beaten
1 onion, chopped
¼ tsp pepper
2 tsps salt
2 tbsps margarine
5 tbsps jam, preferably pineapple
½ cup barbecue sauce *or* 2 tbsps vinegar added to ½ cup tomato sauce, *or* chopped fresh tomatoes stewed with chopped fried onion and ½ clove garlic

Method

1 Combine the meat, breadcrumbs, eggs, onion, pepper and salt. Shape into small balls.
2 Brown the meatballs in hot margarine and put them in a 2 quart casserole.
3 Combine the jam with the barbecue sauce and pour over the meatballs.
5 Bake at 350 °F (180 °C, gas mark 4) for 30 minutes, stirring occasionally.

Note: these can be served as an appetizer or as a main course with spaghetti. If served with spaghetti, use less jam and more tomatoes, vinegar, garlic and onions, and add pepper and a pinch of oregano.

Sweet and Sour Pork   *Julia Gitobu*

Serves 6

Ingredients

750 g boned pork
$\frac{1}{2}$ tsp salt
1 tbsp fat
3 cups water
$\frac{1}{2}$ cup pineapple juice
3 tbsps sugar
3 tbsps cornflour
$\frac{1}{4}$ cup cider vinegar
2 tbsps soy sauce
$\frac{1}{4}$ tsp ground ginger
1 green pepper
1 onion
$1\frac{1}{4}$ cups pineapple chunks

Method

1 Wash the meat and remove unwanted fat. Chop the meat and sprinkle with salt.
2 Heat the fat in a saucepan and brown the meat in the fat.
3 Add $2\frac{1}{4}$ cups of water and simmer for 60 minutes.
4 Combine the pineapple juice with $\frac{3}{4}$ cup water and the sugar, cornflour, vinegar, soy sauce and ginger. Mix well and pour this mixture over the meat. Stir.
5 Cook until slightly thickened (about 5 minutes).
6 Cut the green pepper into strips and slice the onion. Add to the meat and bring to the boil.
7 Add the pineapple chunks. Bring to the boil and simmer for 5 minutes.

Goat's Meat Ngerima

Jimmy Mugo

Ingredients

3⅓ cups water
2 tsps mixed herbs
2 tsps mixed spice
1 tsp chopped chilli pepper
750 g minced goat's meat or beef
¼ cup lemon juice
1 beef stock cube
500 g goat's liver
2 goat's kidneys
¼ cup salted goat's blood
1 goat's rumen

Method

1. Warm 3⅓ cups water and add the mixed herbs, mixed spice, chilli, minced meat and lemon juice.
2. Let it stand overnight or for 3 hours.
3. Simmer the marinated meat for 30 minutes and add the stock cube as it cools.
4. Boil and mince the liver and kidneys and mix with salted blood. Allow it to cool.
5. Clean the rumen and turn it inside out.
6. Mix the cooked meat with the mixture of minced liver, kidney and blood.
7. Stuff the rumen, skewer and boil for 30 minutes. Brown under the grill and collect the dripping. Serve hot, using the dripping as a sauce.

Roast Mutton Ribs

Irene Gitahi

Serves 4

Ingredients

500 g mutton ribs
juice of 1 lemon
2 tbsps fat
1 tsp paprika
salt to taste

Method

1. Wash the mutton ribs and sprinkle with the lemon juice to tenderise them.
2. Cut the ribs apart lengthwise. Smear them with fat and sprinkle with paprika.
3. Pre-heat the oven to 325 °F (160 °C, gas mark 3).
4. Put the ribs on a grilling rack and cook in the oven for 45–60 minutes.
5. Add salt to taste before they are completely cooked. Serve hot.

Grilled Mutton Ribs

Anne Mureria

Serves 4

Ingredients

500 g mutton ribs
1 large lemon
½ tsp black pepper
1 tsp salt

Method

1 Wash the meat and cut the ribs apart along the bone.
2 Squeeze the lemon, smear the juice over the ribs and sprinkle with black pepper.
3 Grease a roasting tin and arrange the pieces in it.
4 Grill or bake the meat in a moderate heat for 30 minutes on each side. (Mutton ribs can also be roasted on a wire rack over a charcoal burner.) Baste the ribs with the extracts.
5 When almost ready sprinkle them with a little salt. Serve hot.

Mau Meat Cakes

Mary Kamami

Serves 3

Ingredients

1 onion
250 g minced beef
¼ tsp salt
pinch of pepper
1 tbsp flour
1 tbsp fat
1 large tomato or 1 large onion to garnish

Method

1 Finely chop or grate the onion.
2 Mix the minced meat with the salt, pepper and chopped onion.
3 Using a little flour, shape the mixture into small cakes.
4 Heat the fat in a frying pan and fry the cakes gently in the fat for 10 minutes each, one at a time, until firm and cooked inside.
5 Drain on absorbent paper and serve hot. Garnish with sliced tomatoes or onion rings.

Lumonde

Sylviah Murungi

Serves 2

Ingredients

360 g beef
1 tbsp fat
1 medium onion
1 tsp curry powder
2 bananas
2 medium potatoes
4 cups water
1 tsp salt

Method

1. Chop the meat and onion. Fry the onion in the fat and add the curry powder. After about 2 minutes, add the meat and fry for 5 minutes.
2. Peel the bananas and potatoes and add them to the meat with the water and salt.
3. Cook for 15-20 minutes, then mash to a thick consistency. Serve hot with vegetables.

Meat and Vegetable Pie

Dorcas Kinuthia

Serves 4

Ingredients

For the shortcrust pastry:
1 tbsp cooking fat
$\frac{1}{4}$ cup margarine
$\frac{3}{4}$ cup flour
$\frac{1}{2}$ tsp salt
4-5 tbsps water
1 tsp lemon juice
1 egg, beaten, or a little milk

For the filling:
250 g stewing steak
1 small onion
1 tbsp salad oil
$1\frac{1}{2}$ cups water
1 cup shelled peas
1 medium carrot
$\frac{1}{2}$ bunch dhania
$\frac{1}{2}$ bunch sukumawiki
$\frac{1}{2}$ cup maize
$\frac{1}{4}$ tsp pepper
$\frac{1}{2}$ tsp salt
1 tomato

Method

1. Cut the meat into small pieces and chop the onion. Fry both in a saucepan until brown. Add $1\frac{1}{2}$ cups water.
2. Add the peas, prepared vegetables and seasonings and simmer for 30 minutes.
3. Meanwhile, make the pastry. Rub the fat into the flour and salt until it is the texture of fine breadcrumbs. Add water and lemon juice and mix to a stiff but easy to roll consistency.
4. Turn the pastry onto a lightly floured board and bring the mixture together lightly with the fingertips. Pre-heat the oven to 425 °F (220 °C, gas mark 7).
5. Roll the pastry to the size and shape of the dish, with $1\frac{1}{2}$ cm extra all round.
6. Put the dish upside down on to the pastry and cut round it, leaving an extra $\frac{1}{2}$ cm.
7. Damp the edge of the dish and line it with pastry. Use the extra pastry to make leaves.
8. Pour the filling into the dish. Damp the pastry edge and cover the pie; decorate the edge with a fork or with your fingers. Arrange the pastry leaves around the top and make a small hole in the middle.
9. Brush with beaten egg or milk.
10. Bake for 20–25 minutes or until golden brown. Garnish with parsley.

Note: the filling can be any meat and vegetable mixture. It should not be too liquid or it will dampen the pastry.

Farmer's Pie
Sarah Chilumo

Serves 2

Ingredients

1 onion
1 tomato
2 tbsps kimbo
250 g minced meat
1 green pepper
1 tsp ground garlic
2 tbsps tomato paste
1 cup water
5 cooking bananas
1 tsp salt
1 egg, beaten

Method

1. Pre-heat the oven to 350 °F (180 °C, gas mark 4).
2. Grease a 600 ml pie dish. Chop the onion finely and scald, skin and chop the tomato.
3. Heat the fat and fry the onion until soft.

4 Fry the minced meat, separating with a fork, for 2 minutes.
5 Add the tomato, chopped sweet pepper, ground garlic and tomato paste.
6 Add the water and simmer for 5 minutes.
7 Turn the mixture into a pie dish and bake for 30 minutes.
9 Peel, wash, boil and mash the bananas.
10 Cover the pie with the mashed bananas and spread beaten egg on top. Bake for 15 minutes or until firm and brown on top. Serve hot.

Stuffed Green Peppers
Gacheri Mburugu

Serves 4

Ingredients

½ cup rice
250 g minced meat
2 tbsps margarine
2 medium tomatoes
1 medium onion
½ tsp curry powder
pinch of ground ginger
pinch of black pepper
pinch of salt
4 large green peppers

Method

1 Boil the rice in salted water.
2 Fry the minced meat in margarine, add the chopped tomatoes, onion and spices and cook for 10 minutes.
3 Mix the cooked rice with the cooked meat.
4 Cut a hole in the top of each green pepper and remove the seeds.
5 Fill each green pepper with the mixture of meat and rice.
6 Bake in a casserole for about 20 minutes, or until the green peppers are soft, at 375 °F (190 °C, gas mark 5). Serve hot.

Potato-Meat Roll
Anne Mwangi

Serves 4

Ingredients

4 large potatoes
1 cup milk
2 tbsps margarine
2½ cups flour
2 tbsps fat

250 g minced meat
1   sweet pepper
1   onion
1   tbsp chopped dhania
2   eggs, beaten

Method

1 Boil the potatoes and mash them with the milk, flour and margarine.
2 Fry the minced meat and add the chopped sweet pepper, onion and dhania.
3 Roll out the potato pastry on a floured board.
4 Put the cooked minced meat in the centre and roll the pastry round it.
5 Shape the pastry into an oblong and pinch the ends together. Brush the top with beaten egg and cook in a moderate oven (350 °F, 180 °C, gas mark 4) for 20–30 minutes.

Offal

The types of offal (the internal organs of an animal) most commonly eaten are liver, kidneys and tripe (the lining of ox stomachs). Blood is eaten by many people in Kenya, on its own or in combination with meat. Liver and kidneys are excellent sources of protein, iron and vitamins A and B, and tripe is rich in calcium.

Offal should always be washed thoroughly before being prepared. The outer skin of liver and any blood vessels should be removed. Kidneys should be skinned and then cored by cutting from the round side and removing the core with kitchen scissors. Soaking kidneys in milk before cooking them (the milk should then be thrown away) helps to improve their flavour.

Liver and kidneys can be lightly fried or grilled. Tripe, because it has a bland flavour, is best stewed with strong-flavoured ingredients.

Mutura
(Kikuyu Sausage)

Irene Gitahi

Serves 4

Ingredients

½ metre large intestine (colon)
250 g small intestine (duodenum)
1 kidney
250 g liver
1 medium onion
1 tbsp fat
1 tsp paprika
1 tsp black pepper
2 cups blood
salt to taste

Method

1 Clean the large and small intestines thoroughly.
2 Finely chop the small intestine, kidney and liver. Slice the onion and fry it in hot fat.
3 Add all the chopped meat and the paprika and pepper. Simmer for 30 minutes.

4 Add the blood and salt to taste. Cook until tender (about 20 minutes).
5 Stuff the fried meats and blood into the large intestine and tie both ends. Bake or grill for 20–25 minutes at 375 °F (190 °C, gas mark 5).

Note: this dish is best made with goat's meat.

Sautéed Liver

Ruth Oniang'o

Serves 2

Ingredients

2 medium onions
2 tbsps margarine
2 tbsps cooking fat
250 g liver
4 tbsps flour
$\frac{1}{2}$ tsp salt

Method

1 Slice the onions into rings and sauté until tender. Remove from the fat.
2 Clean and dry the liver and cut it into slices about $\frac{1}{2}$ cm thick. Dust with flour and salt.
3 Sauté the liver for about 5 minutes or until no blood comes out when pricked with a fork.
4 Lift the slices of liver and place the onion rings underneath them.
5 Cover the pan and cook on a low heat until the onions are tender (about 10 minutes).

Spiced Coconut Liver

Magdalen Juma

Serves 2

Ingredients

250 g liver
$\frac{1}{4}$ tsp turmeric powder
salt to taste
$\frac{1}{2}$ fresh green chilli pepper, finely chopped
$\frac{1}{4}$ tsp ground ginger
1 onion
$\frac{1}{2}$ tomato
1 tbsp cooking fat
$\frac{1}{2}$ tsp tomato purée
$1\frac{1}{2}$ cups coconut milk

Method

1 Clean and chop the liver into cubes. Mix the spices together and rub into the pieces of liver.
2 Skin and chop the onion and tomato. Fry the onion in hot fat until it is soft. Add the liver, tomato, tomato purée and coconut milk and simmer for 1 hour. Serve hot with rice, mashed bananas, chapatis or mashed potatoes.

Liver Omelette   *Florence Muthoni*

Serves 2

Ingredients

1 onion
50 g liver
1 tbsp kimbo
2 eggs
1 tbsp milk
$\frac{1}{2}$ tsp salt

Method

1 Finely chop the onion and liver.
2 Fry the onion in half the kimbo until golden brown. Add the liver and cook for 2 minutes. Remove and keep aside.
3 Beat the eggs with the milk and salt.
4 Add the cooked liver and onion to the egg mixture and mix well.
5 Heat the rest of the kimbo in an omelette pan. Add the egg and liver mixture and stir until it begins to set. Allow to brown slightly and serve immediately.

Liver with Yoghurt   *Veronica Onyango*

Serves 3

Ingredients

350 g liver
$\frac{1}{2}$ tbsp flour
$\frac{1}{4}$ tsp salt
$\frac{1}{4}$ tsp pepper
2 onions
2 tomatoes
1 tbsp fat
2 tbsps tomato purée

1 tsp Worcester sauce
142 g carton yoghurt
1 tsp gravy browning
parsley to garnish

Method

1 Cut the liver into strips and toss lightly in seasoned flour. Slice the onions. Skin and halve the tomatoes, remove the seeds and chop.
2 Melt the fat in a heavy frying pan and lightly brown the onion.
3 Add the liver and fry to seal on all sides.
4 Stir in the rest of the ingredients except the gravy browning. Bring to the boil, stirring, and simmer for 5 minutes.
5 Stir in the gravy browning. Serve in a shallow dish, garnished with parsley.

Matoke Tripe
Irene Gitahi

Serve 2

Ingredients

250 g tripe (matumbo)
5 potatoes
3 green bananas
1 small onion
2 tbsps fat
salt to taste
1 tsp curry powder

Method

1 Clean the tripe thoroughly and chop into small pieces. Boil for 30 minutes.
2 Prepare and chop the potatoes and bananas. Slice the onion and fry until soft. Add the tripe and cook for 60 minutes.
3 Add potatoes, bananas, salt, curry powder and $2\frac{1}{2}$ cups water. Cook until tender (about 30 minutes) and serve hot.

Poultry

Chicken is the most popular type of poultry, but the term also includes turkeys, ducks, geese and guinea fowl. Poultry is widely available in Kenya, and keeping domestic fowl is becoming increasingly popular. Poultry is a good source of easily digested protein and some vitamin B (niacin).

Plucking and drawing poultry

1. Unless the bird is very young, plunge it head down into boiling water. This makes the feathers easier to remove, and will also kill any insects on the bird.
2. Pull out the feathers quickly, against the way they lie.
3. Singe off any hairs over a smokeless fire or with a wooden taper.
4. Make a slit in the inside of each leg below the 'knee' and pull out the leg sinews.
5. Cut off the legs about an inch below the 'knees' and dip the stumps in boiling water for about 10 seconds.
6. Cut off the neck as close to the body as possible. Make a 2 cm slit from the head end down the back and draw out the windpipe and crop, taking care not to break the crop.
7. Make a small horizontal slit just above the anus. Insert two fingers and carefully draw out the internal organs. Be careful to remove everything. The giblets – the gizzard, heart and liver – may be kept to make gravy.
8. Cut away the skin around the anus.
9. Thoroughly wash and dry the bird inside and out.

If the bird is bought already dressed, it should be washed before use.

Chicken Stew *Maureen Owuor*

Serves 4

Ingredients

1 tbsp kimbo
1 onion
250 g chopped chicken
3 carrots
3 tomatoes

1 bunch dhania
¾ tsp salt
1 tsp curry powder
100 g chopped beef liver
1 tbsp flour

Method

1. Melt the fat until it bubbles gently and fry the chopped onion until transparent.
2. Brown the chicken gently in the same pan.
3. Add chopped vegetables, dhania, salt, curry powder and 1 cup water.
4. Cover with a tight-fitting lid and cook gently until done (35-60 minutes, depending on the tenderness of the chicken). Add the liver and cook for 10 minutes.
5. Blend the flour with a little water in a cup. Take the pan off the heat and stir in the paste to thicken the gravy. Let it simmer for a few minutes.
6. Serve hot with vegetables.

Spiced Chicken in Yoghurt

Diana Barwa

Serves 4

Ingredients

1 chicken (about 1 kg)
1 tbsp flour
⅓ cup butter
3 medium onions
1 clove garlic
2 tsps ginger
2 tsps paprika
2 bay leaves
pepper and salt to taste
4 cloves
1¼ cups water
⅔ cup yoghurt

Method

1. Joint the chicken and toss the pieces in flour. Melt the butter and lightly brown the chicken.
2. Chop the onions and garlic. Remove the chicken pieces from the pan and add the onions, garlic, spices and seasoning to the fat. Heat gently until the onion is soft.
3. Return the chicken to the pan and add the water. Cover and simmer for 20 minutes. Stir in the yoghurt just before serving.

Note: this recipe is very popular at the Coast.

Chicken Special
Alice Kirigia

Serves 4

Ingredients

half a 2 kg chicken
1 large onion
2 carrots
½ bunch dhania
2 tomatoes
1 large green pepper
2 tbsps fat
¼ tsp salt
pinch of pepper
1 chicken stock cube

Method

1 Skin, bone and chop the half chicken. Peel and chop the onion. Clean and grate the carrots and chop the dhania, tomatoes and green pepper.
2 Fry the onion in hot fat for a few minutes without letting it brown.
3 Add the chicken, cover and cook on a low heat until soft. Drain off the excess fat.
4 Add carrots, green pepper, tomatoes and dhania. Cook for 10–15 minutes in a little water.
5 Add salt, pepper and stock made with the stock cube and 3 cups of warm water. Simmer for 10 minutes. Serve with rice, chapatis or mashed potatoes.

Simmered Chicken
Martha Nyangi

Serves 3–4

Ingredients

half a 1 kg chicken
1 medium onion
2 medium tomatoes
1 bunch dhania
1 large green pepper
1 tbsp kimbo
¼ tsp curry powder
1½ cups water
1 tsp salt

Method

1. Joint and skin the half chicken. Chop the onion, tomatoes, dhania and green pepper.
2. Fry the onion until golden brown. Add chicken pieces, chopped tomato, dhania, sweet pepper and curry powder. Cook on a low heat for 10 minutes.
3. Add the water and bring to the boil. Simmer for 1 hour. Add salt 10 minutes before removing from heat. Serve with rice, ugali, irio or chapatis.

Chicken Paprika
Rosemary Mutunga

Serves 4

Ingredients

500 g boned chicken
2 tbsps fat
1 small onion
3 medium carrots
3 cups hot water
1 chicken stock cube
1 tsp paprika
salt to taste
1 tbsp flour

Method

1. Chop the chicken, brown the pieces in hot fat and put to one side.
2. Chop the onion and carrots. Fry the onion in the fat until soft.
3. Add the hot water to the stock cube
4. Add the chicken to the cooked onion, then add paprika, salt, flour and stock. Simmer for 30–40 minutes.
5. Add the carrots before completely cooked. Serve hot.

Kenya Chicken
Jane Kibuga

Serves 4

Ingredients

1 small chicken
½ cup grated cheese
½ bunch parsley
¾ cup breadcrumbs
1 tsp white pepper
1 tsp salt
1 cup margarine

Method

1. Joint the chicken.
2. Mix the grated cheese, chopped parsley, breadcrumbs, pepper and salt in a shallow plate.
3. Melt the margarine and dip the chicken pieces into the margarine, then into the breadcrumb mixture so that each piece is covered.
4. Bake in a greased tin for 45 minutes at 325 °F (160 °C, gas mark 3).

Chicken Brochette Coastal Style *Charles Ogutu Ohomo*

Serves 4

Ingredients

1 roasting chicken
½ cup lemon juice
½ cup sour cream
salt and pepper to taste
a little soy sauce
a little chopped dhania
a little coconut milk
1 green pepper
1 onion
1 tomato

Method

1. Bone the chicken and chop into large cubes.
2. Marinate the chicken in the lemon juice, sour cream, salt, pepper, soy sauce, dhania and coconut milk for about 30 minutes in a cold place.
3. Quarter the green pepper, onion and tomato. Skewer the pieces of chicken, alternating with the pepper, onion and tomato. Grill over a low heat for 10 minutes, basting from time to time.
4. Serve with rice, chopped coconut and dhania.

Dhania Chicken *Charles Ogutu Ohomo*

Serves 4

Ingredients

1 small capon
1 tbsp oil
3 cloves garlic, crushed
1 tsp ground garlic
1 onion, chopped

1 bunch dhania, chopped
2 tbsps tomato purée
salt and pepper to taste

Method

1 Joint the chicken. Heat the oil in a frying pan and fry the chicken until golden brown.
2 Pour off the fat and add the rest of the ingredients. Cook on a low heat for 15 minutes.

Pakistan Chicken Curry *Leah Marangu*

Serves 6

Ingredients

1 large onion, sliced
2 tbsps kimbo
3 cups water
pinch of salt
$\frac{1}{4}$ tsp ground coriander
$\frac{1}{8}$ tsp ground ginger
$\frac{1}{4}$ tsp ground turmeric
1 bay leaf
$\frac{1}{4}$ tsp garlic salt
$\frac{1}{8}$ tsp cayenne
2 kg chicken pieces
1 cup sour cream

Method

1 Sauté the onion in half the kimbo until golden brown.
2 Add 1 cup of water and all the spices and boil for 10 minutes.
3 Brown the chicken in the rest of the kimbo. Add 2 cups of water and the onion and spice mixture.
4 Simmer until the chicken is tender. Add the sour cream just before serving. Serve with rice.

Fish

Fish can be divided into two main groups: freshwater fish (e.g. tilapia, catfish and dagaa) and sea fish (e.g. cod, mackerel, salmon, haddock and shark). Fish is a good source of protein, calcium and phosphorous, and sea fish contains iodine. It contains very little carbohydrate and its fat content ranges from 2% to 10%. The fat content of cooked fish will depend on the method of preparation: fried fish, for example, will absorb fat. Fat or oily fish contains more vitamins (especially the fat-soluble vitamins found in fish livers) than lean or white fish. Fish has a fair supply of B group vitamins.

Preparing fish for cooking

1. Hold the fish by the tail and remove the scales by scraping with the back of a knife from the tail to the head. Wash off any loose scales.
2. Cut off the fins and tail. Cut off the head just below the gills.
3. Make a slit down the whole of the belly, remove all the entrails, scrape out the inside of the fish and wash well.

Fish can be boiled, steamed, roasted, grilled or baked. It is highly perishable, so as little time as possible should elapse between catching or buying and eating it.

Tilapia in Coconut Milk *Magdalen Juma*

Serves 2

Ingredients

1 tilapia
1 onion
1 tomato
$\frac{1}{2}$ tsp cooking fat
pinch of salt
1 cup coconut milk
$\frac{1}{2}$ tsp tomato purée
1 lemon

Method

1. Prepare and scale the fish.
2. Chop the onion, wash and slice the tomato.

3 Melt the fat and fry the onion and tomato until the onion is soft. Add the fish, salt, coconut milk and tomato purée.
4 Cook on a low heat for 30 minutes. Serve with rice, ugali, mashed bananas or potatoes and wedges of lemon.

Variation: add $\frac{1}{2}$ tsp mixed spice with the onion.

Stewed Tilapia

Damari Otieno

Serves 1

Ingredients

1 tbsp fat
1 medium onion, chopped
1 tomato
$1\frac{1}{2}$ cups water
1 medium tilapia
1 tsp lemon juice
$\frac{1}{2}$ tsp black pepper
$\frac{1}{3}$ tsp ground turmeric
$\frac{1}{2}$ tsp salt
1 lemon

Method

1 Melt the fat and fry the onion until golden brown.
2 Add the skinned and chopped tomato and fry for 2 minutes, stirring continuously.
3 Add the water and bring to the boil.
4 Add the cleaned and dressed fish and simmer for 15 minutes.
5 After 10 minutes, add the lemon juice, pepper, turmeric and salt.
6 Serve with ugali garnished with slices of lemon.

Tilapia in Mrenda

Pauline Owiti

Serves 1

Ingredients

2 bunches mrenda
1 medium onion
1 medium tomato
1 medium tilapia, slightly dried
1 cup water
pinch of bicarbonate of soda
2 tbsps cooking fat

¼ tsp curry powder
½ cup fresh milk
salt to taste

Method

1. Prepare and chop the mrenda, onion and tomato.
2. Wash the fish thoroughly in warm water.
3. Bring 1 cup of water to the boil and add a pinch of bicarbonate of soda.
4. Add the mrenda and cook for 2–3 minutes, stirring continuously.
5. Remove from the heat and put aside.
6. Heat the cooking fat, add the chopped onion and curry powder and fry until the onion is golden brown.
7. Add the milk, mrenda and tomato, stir well and add the fish. Cover the pan and simmer gently for 30–40 minutes or until tender.
8. Add salt to taste and cook for 5–7 minutes. Serve with ugali or rice.

Tilapia in Peanut Sauce *Wilfrida Amenya*

Serves 4

Ingredients

1 large tomato
1 small onion
2 medium tilapia
1 tbsp fat
½ tsp salt
½ tsp black pepper
1 tbsp tomato purée
1⅔ cups water
2 tbsps peanut butter
sliced lemon to garnish

Method

1. Skin and dice the tomato and onion.
2. Scale, clean and bone the fish.
3. Melt the fat, add the onion and cook until tender.
4. Add the tomato and cook until soft.
5. Add seasonings, tomato purée, 1 cup of water and the fish. Bring to the boil and simmer for 20–30 minutes.
6. Mix the peanut butter with the rest of the water.
7. When the fish is cooked, add the peanut sauce.
8. Serve hot with ugali, garnished with sliced lemon.

Spicy Tilapia

Mary Mulaku

Serves 2

Ingredients

1 onion
1 tbsp fat
1 small bunch dhania
1 medium carrot
1 tomato
1 tsp curry powder
pinch of mixed spice
pinch of white pepper
salt to taste
½ cup groundnuts
1 tsp roiko mchuzi mix
1 small tilapia

Method

1. Sauté the chopped onion in hot fat until brown.
2. Add chopped dhania, grated carrot and chopped tomato and fry until tender.
3. Add the spices and salt. Pound the groundnuts and add them to the mixture with the roiko mchuzi mix.
4. Add the scaled and salted fish and turn carefully in the mixture without breaking.
5. Cook at medium heat for 7–10 minutes. Halve the tilapia and serve sizzling hot with matoke, ugali or sweet potatoes.

Omena in Coconut Milk

Lynette Opondo

Serves 2

Ingredients

2 cups omena
1 onion
2 tomatoes
1 tbsp fat
½ tsp curry powder
1 tbsp tomato purée
1 cup coconut milk
½ tsp salt

Method

1. Wash the omena in warm water. Chop the onion and tomatoes.
2. Heat the fat and fry the onion and tomatoes until tender. Add curry powder and tomato purée.

3 Add the omena and fry for 2 minutes.
4 Add the coconut milk and salt and simmer for 20 minutes. Serve hot with ugali.

Dried or Smoked Ngege Stew *Leunita Muruli*

Serves 4

Ingredients

2 medium dried or smoked ngege
1 medium onion
2 tsps salt
2 medium tomatoes
1 cup boiled milk

Method

1 Wash the fish. Put them in a saucepan with the chopped onion, salt and just enough water to cover the fish.
2 Cook for one hour at medium temperature with the pan open, stirring gently from time to time.
3 Slice the tomatoes, add to the fish and cook for 5 minutes.
4 Remove from the heat and gently stir in the milk. Season to taste and serve hot with ugali.

Seafish with Coconut and Ginger *Charles Ogutu Ohomo*

Serves 4

Ingredients

1 large whole fresh seafish
1 cup melted butter
1 tbsp chopped tarragon
1 cup chopped onions
8 cloves
2 tbsps grated fresh ginger
1 cup grated coconut
salt and pepper to taste

Method

1 Clean, gut and scale the fish and put it in a greased roasting tin.
2 Smear the inside of the fish with half the melted butter and stuff it with the tarragon and onions.

3 Stud the top of the fish with cloves and cover with grated ginger and coconut. Coat the fish with butter, season to taste and bake at 375 °F (190 °C, gas mark 5) for 40–50 minutes.

Fried Fish Fillets

Anne Masenge

Serves 3

Ingredients

3 fish fillets
½ cup flour
pinch of curry powder
½ tsp salt
1 egg
1 cup breadcrumbs
1 tbsp cooking fat

Method

1 Wash and dry the fish fillets and toss them in the flour mixed with the curry powder and salt.
2 Beat the egg lightly. Dip each fillet in the beaten egg, then coat it with breadcrumbs.
3 Fry in very little fat, turning frequently, until brown.
4 Drain the fish on absorbent paper and serve with lemon sections.

Fried Fish and Groundnuts

Joy Awori

Serves 2

Ingredients

1 medium-sized dried fish
1½ cups groundnuts, pounded
salt to taste
1 onion

Method

1 Boil the fish until cooked, remove from the stock and add enough water to the stock to make 3 cups.
2 Add the groundnuts to the water and cook, stirring the mixture, until fat forms on top. If the mixture becomes too thick, add more water.
3 Add the salt and chopped onion. Stir and add the cooked fish. Boil again before serving. Serve with matoke or rice.

Fishcakes

Magdalen Musila

Makes 4 fishcakes

Ingredients

1 cup cooked fish flakes
5 medium potatoes, boiled and mashed
pinch of salt
1 egg, beaten with 1 tbsp water
1 cup breadcrumbs
3 tbsps cooking fat

Method

1. Mix together the fish, potatoes and salt and mould into cakes about 4 cm in diameter.
2. Dip the fishcakes first into the egg, then into the breadcrumbs.
3. Fry in hot fat until golden brown on both sides, turning once. Serve hot with tomato sauce.

Lamu Coconut Prawns

Charles Ogutu Ohomo

Serves 4

Ingredients

1 coconut
2 tbsps butter
1 kg unpeeled prawns
1 cup chopped onions
2 tbsps chopped mushrooms
1 cup chopped tomatoes
soy sauce to taste
white wine to taste
salt and pepper to taste

Method

1. Split the coconut in half, remove the flesh and cut it into small strips. Heat the butter and sauté the coconut until golden brown. Remove from the pan and set aside.
2. Peel and wash the prawns and sauté with the onion for 3 minutes. Add the remaining ingredients and cook for 5 minutes. Season and serve with rice.

Cereal-based dishes

Cereals are traditional foods in Kenya, eaten regularly and in large quantities. They are grown locally, are cheap and easy to transport and store. They supply the bulk of the energy needs of most people. Maize and millet are staple foods in Kenya. Maize is rich in carbohydrates and contains small amounts of poor quality protein which is deficient in the essential amino acids tryptophan and lysine. Millets and sorghum are regaining their popularity and are widely used in the preparation of porridge. Various types of both are found in Kenya: finger millet, bulrush millet and red and white sorghums. They can be milled to make flour and are good sources of carbohydrate, protein, calcium and vitamin E.

Ugali is a popular Kenyan staple which is made from flours of different cereals, the most popular being maize flour. Ugali made from wimbi and sorghum flours is usually a little more difficult to make because of the stickiness of the flours, especially when cassava has been added to them. Continuous stirring will prevent lumps forming, and the flour must boil, absorb water and swell.

Wimbi and Maize Ugali *Miriam Sei*

Serves 1–2

Ingredients

½ cup wimbi flour
1 cup maize flour
2 cups water
½ tsp margarine or butter
pinch of salt

Method

1. Mix and sift the wimbi and maize flour.
2. Boil the water and add the margarine and salt.
3. Add the flour to the boiling water. Stir the mixture until thick, then cover and simmer for 5 minutes.
4. Stir the mixture again, and keep stirring until the ugali rolls on the pot when turned. Serve hot with stewed meat, roast meat or fish, green vegetables and fresh or sour milk.

Enriched Ugali

Leah Kibiego

Serves 1

Ingredients

½ cup water
½ cup sour milk
1 tbsp margarine
½ tsp salt
1 cup wimbi flour

Method

1. Boil the water, add the sour milk and boil for 2 minutes.
2. Add the margarine and salt.
3. Add the flour gradually and cook until it reaches a thick, smooth consistency. Serve hot with a stew.

Carrot Rice

Lilian Muthuuri

Serves 2

Ingredients

2 carrots
2 tomatoes
1 onion
2 tbsps cooking fat
½ tsp curry powder (optional)
1 cup rice
salt to taste

Method

1. Peel and grate the carrots. Blanch the tomatoes and chop them finely. Peel and chop the onion.
2. Melt the fat and fry the onion until soft. Add the diced tomatoes, curry powder and rice and mix well.
3. Pour on enough boiling water (about 2 cups) to cook the rice. Add salt to taste.
4. Boil until the rice is soft, then add the grated carrots and lower the heat. Cook for 5–10 minutes, or until the rice is cooked and the water has been absorbed. Serve hot with meatballs, Scotch eggs or meat stew.

Cereal-based dishes 83

Green Peas and Coconut Rice
Florine Mbithi

Serves 2

Ingredients

¾ cup green peas
1 cup rice
2 cups coconut milk
salt to taste

Method

1 Boil the peas and clean and wash the rice.
2 Bring the coconut milk to the boil. Add the salt and rice and cook for 25 minutes, stirring occasionally.
3 Add the cooked green peas and mix thoroughly. Cook until the coconut milk has dried out, then put into a fireproof dish, cover and cook in the oven for 15 minutes at 375 °F (190 °C, gas mark 5). Serve with meat or chicken stew.

Spiced Rice
Leah Githinji

Serves 2

Ingredients

¼ medium onion
½ bunch dhania
1 chilli pepper
1 tbsp kimbo
1 tsp cumin seeds
1 tsp mixed spice
1 tsp beef roiko mchuzi mix
1 cup rice
2½ cups water

Method

1 Chop the onion, dhania and chilli pepper, fry them in the kimbo and add the cumin seeds, mixed spice and roiko mix.
3 Wash the rice and add it to the mixture.
4 Add the water and cook until soft. Serve with meat stew.

Variation: add 1 diced carrot and 500 g chopped meat at the same time as the rice.

Flaky Chapatis

Janet Ngoci

Makes 8 chapatis

Ingredients

3½ cups sifted flour
½ tsp salt
1 tbsp margarine
1⅓ cups warm water
½ cup oil

Method

1. Mix together the flour and salt and rub in the fat, using the fingertips, until all the fat is well mixed in.
2. Make a well in the centre and pour in half the water. Mix with a wooden spoon, adding a little water at a time, until the dough is firm but soft. Continue mixing with your hands.
3. Knead the dough and divide into eight balls. Roll each ball into a circle, rub the top with oil and fold it into a wheel.
4. Roll out each wheel into a circle 2 cm thick.
5. Fry each circle on a low heat on both sides until golden brown. Keep the chapatis warm until ready to serve.

Dhania Chapatis

Wilfrida Amenya

Makes 4 chapatis

Ingredients

2 cups plain flour
¾ tsp salt
1 bunch dhania
1 cup hot milk
2 tbsps margarine
½ cup oil

Method

1. Sift the flour and salt together and add the finely chopped dhania.
2. Add the hot milk to make a soft dough.
3. Knead in the margarine.
4. Knead further on a floured board until the dough does not stick to the hands. Shape into four small balls about 3 cm in diameter.
5. Roll out and brush a little oil on to each ball.
6. Fold into a roll and shape into a ring.
7. Leave to rest for about 10 minutes.
8. Roll out again into a circle and cook on an ungreased pan. Serve with any stew.

Cereal-based dishes 85

Coconut Milk Chapatis

Lynette Oponda

Makes 4 chapatis

Ingredients

2 cups flour
½ tsp salt
¾ cup coconut milk
⅛ cup salad oil

Method

1. Sieve the flour into a mixing bowl.
2. Add the salt to the coconut milk and add to the flour. Knead thoroughly until the dough is soft. Divide the dough into four balls.
3. Using a rolling board and pin, roll out each chapati, brush a little oil on it and coil it round your finger.
4. Roll each chapati into a circle. Brush oil on one side, fry and repeat with the other side.
5. Serve hot with any stew and/or greengram sauce.

Mukoomo

Margaret Ngau

Serves 2

Ingredients

1 cup soft green maize
2 cups shelled peas
1 cup washed and chopped cowpea leaves
1 cup peeled and cubed young pumpkin
1 tsp salt
1 tsp margarine

Method

1. Clean and wash the maize and peas. Put them, with the cowpea leaves, in a medium-sized saucepan half filled with water and boil, uncovered, for 20–30 minutes.
2. Add the prepared pumpkin, cover the pan tightly and cook for 20–30 minutes or until soft. Remove from the heat and drain off the stock into a bowl.
3. Add salt and margarine and mash, adding stock as required. Serve hot.

Mashed Green Maize and Peas

Sophie Kirorei

Serves 8

Ingredients

4 cups green maize
6 cups shelled green peas
5 large potatoes
2 heaped tbsps margarine
pinch of salt

Method

1 Boil the maize and peas separately until tender.
2 Peel and chop the potatoes.
3 Cook the maize, peas, potatoes and salt together until the potatoes are cooked.
4 Mash with the margarine and serve hot with meat stew.

Madio

Rozinah Kiwinga

Serves 4

Ingredients

1 cup green maize
1½ cups shelled green peas
1 tsp salt
1 medium onion
1 tbsp fat
125 g minced meat
1 fresh tomato
1 tsp chopped chilli pepper
1 carrot (optional)
1 arrowroot
1½ cups water
1½ cups coconut milk
grated carrot to garnish

Method

1 Boil the maize and peas together in salted water until tender.
2 Slice the onion and fry until tender but not brown.
3 Add the meat and cook for 2 minutes. Add the blanched and chopped tomato and the chilli. Add 1½ cups water and cook for 30 minutes or until the meat is tender. (A chopped carrot may be added at this stage.)

4 Prepare and chop the arrowroot into small cubes.
5 Add the arrowroot, maize, peas and coconut milk (utii ya nazi) to the mixture. Cover and cook until the arrowroot is tender and the water has almost evaporated.
6 Add the meat and mash with a wooden spoon.
7 Serve hot, garnished with grated carrot.

Fried Muthokoi
Rosemary Mbithe

Serves 4

Ingredients

1 cup pigeon peas
2 cups muthokoi
1 finely chopped onion
2 tbsps ghee
2 tomatoes, finely chopped
2 carrots, finely chopped
$\frac{1}{2}$ tbsp roiko mchuzi mix
stock cube (optional)
curry powder (optional)
cayenne pepper (optional)
salt to taste

Method

1 Wash the pigeon peas, cover them with water and boil for 30 minutes. Add water as necessary.
2 Clean and wash the muthokoi and add to the semi-cooked pigeon peas. Cook until tender (about 90 minutes).
3 Fry the onion in the ghee until golden brown.
4 Add the tomatoes and carrots and semi-cook them.
5 Add the mchuzi mix, curry powder and stock cube dissolved in water. Stir. Add the pigeon peas, muthokoi, salt and cayenne pepper to taste. Mix well.
6 Cook for 3–5 minutes. Serve hot.

Muthokoi with Meat
Adela Kimeu

Serves 2

Ingredients

1 cup dried beans
$\frac{3}{4}$ cup maize (muthokoi)
2 tbsps fat

1 large onion
125 g meat
1 medium tomato
1 medium carrot
1 large potato
½ tbsp beef roiko mchuzi mix

Method

1. Soak the beans overnight or boil them for 2 hours.
2. Pressure-cook the maize and beans for 45 minutes or until tender.
3. Melt the fat in a saucepan, add the diced onion and fry gently until tender.
4. Add the chopped meat and fry until it changes colour. Add the chopped tomato, carrot, potato and roiko mchuzi, stirring vigorously. Season, add a little water, cover and cook for about 1 hour.
5. Add muthokoi to the stew and cook for about 1 hour. Serve hot.

Muthokoi with Cowpeas

Rosina Mutia

Serves 2

Ingredients

½ cup muthokoi
1 cup cowpeas
1 onion, chopped
1 tbsp fat
1 tsp curry powder
1 tbsp beef roiko mchuzi mix
3 large tomatoes, chopped
250 g meat, chopped
3 carrots
1 cup coconut milk
1 tsp salt

Method

1. Wash muthokoi and cowpeas and pressure-cook them for 1–1½ hours.
2. Fry the onion in a saucepan until golden brown. Add curry powder, roiko mchuzi and tomatoes.
3. Add meat and carrots and cook for 1 hour. Add the coconut milk and bring to the boil. Add the salt.
4. Add the muthokoi to the stew and let it boil. Serve hot.

Meru Muthikore

Alice Kirigia

Serves 2

Ingredients

½ cup pounded maize (muthokoi)
¼ cup dried beans
1 medium onion, sliced
1 tbsp fat
50 g steak, chopped
1 medium carrot, grated
¼ bunch young sukumawiki, sliced
1 large potato, chopped
1 tomato, sliced

Method

1. Soak the maize and beans in hot water for 2 hours.
2. Pressure-cook the maize and beans until tender (about 15 minutes).
3. Fry the onion in hot fat until light brown. Add the meat, stir and cover. Cook on a medium heat until the meat is tender.
4. Add all the vegetables, except the tomato and potato, and cook for 3 minutes. Add the potato and cook for 5 minutes. Add the tomato, stir and season.
5. Add enough water to cover and allow to simmer for 15 minutes or until the potatoes are cooked. Serve hot as a main dish.

Njugu
(Food for in-laws)

Mary Kamami

Serves 2

Ingredients

¾ cup pigeon peas
1 cup green maize
5 potatoes
salt to taste

Method

1. Wash the pigeon peas and soak them overnight.
2. Mix the maize and peas together and cook for 30 minutes in a pressure cooker, or for about 1¼ hours in a saucepan, until soft.
3. Peel the potatoes and add them to the maize and peas. Cook for a further 10–15 minutes.
4. Mash and season with salt. Serve hot.

Maize and Beans
(Amahenjela)

Rita Shirato

Serves 2

Ingredients

1 cup dried beans
½ cup half-dry green maize
2 tbsp cooking fat
1 medium onion
1 medium tomato
salt to taste
1 level tsp curry powder
sliced tomato or grated carrot to garnish

Method

1. Soak the beans overnight.
2. Pressure-cook the maize for 10 minutes, then add the beans and enough water to cover the contents. Pressure-cook for a further 8–10 minutes.
3. Melt the fat and fry the chopped onion until tender.
4. Add the mixture of maize and beans and the chopped tomato, salt and curry powder. Cook for 5 minutes.
5. Serve garnished with tomato or grated carrot.

Maize and Cowpeas

Ruth Oniang'o

Serves 2

Ingredients

1 cup dried cowpeas
1 cup half-dry maize
1 medium onion
2 tbsps cooking fat
½ tsp salt

Method

1. Pressure-cook the cowpeas and maize for 30 minutes at high pressure. Drain and save any excess water.
2. Sauté the chopped onion in fat and add the maize and cowpea mixture with 1 cup of the reserved water and salt to taste.
3. Cook until all the water has been absorbed. Serve hot with strong tea.

Maize and Beans in Sauce

Ruth Oniang'o

Serves 4

Ingredients

1 cup dried kidney beans
1 cup half-dry maize
1 small onion
1 tbsp fat
½ tsp curry powder
½ beef stock cube
½ tsp salt
2 tbsps flour

Method

1. Soak the beans overnight in cold water.
2. Cook the maize and beans in a pressure cooker for 25 minutes. Drain and save the excess water.
3. Chop the onion and sauté in fat, then add the curry powder and cook for 1 minute. Add the maize and beans.
4. Dissolve the stock cube in ½ cup saved water and add to the mixture with the salt.
5. Mix flour in a little water to make a paste and stir this into the maize and beans, without breaking the beans.
6. Cook for 5 minutes longer. Serve hot.

Nyani

Julia Gitobu

Serves 4

Ingredients

1½ cups maize (fresh but not too soft)
2 cups shelled green peas
1 cup finely chopped cowpea leaves
4 large potatoes
¾ tsp salt
1 tbsp butter or margarine (optional)

Method

1. Wash the maize, put it in a medium-sized saucepan or sufuria, cover with water and cook until soft.
2. Add the peas and boil for about 10 minutes.
3. Add the cowpea leaves and continue cooking for about 20 minutes.
4. Add the peeled and sliced potatoes and bananas. Check that the level of the water is about halfway up the pan. Add salt, cover and leave to cook gently for about 1 hour or until the ingredients are soft.

5 Drain off the water and mash. If necessary, soften with drained stock or boiling water. Add margarine or butter to taste and mix well. May be served hot or cold.

Isandi
Veronica Mutua

Serves 6

Ingredients

1 cup millet (mawere)
3 cups water
½ tsp salt
2 cups maize flour

Method

1 Boil the mawere in salted water until the grains open.
2 Thicken with the maize flour as for ngali. Serve with sour milk or meat stew.
Variation: add 1 tbsp margarine or other fat before removing from the heat.

Minced Meat Rice
Leah Marangu

Serves 4

Ingredients

1 medium onion, chopped
1 tbsp margarine
1½ tsp salt
½ tsp white pepper
2 tsps curry powder
3 tomatoes, finely chopped
1 cup minced meat
1 beef stock cube
6 cups water
2 cups rice
1 bunch spinach, chopped

Method

1 Fry the onion in margarine until tender.
2 Add the salt, pepper, curry powder and tomatoes, and cook for 5 minutes, stirring continuously.
3 Add the minced meat and cook for 5 minutes on a low heat.

4 Dissolve the beef stock cube in a little water and add to the meat. Add the rest of the water and bring to the boil.
5 Add the rice and reduce to medium heat. Cook for 10 minutes.
6 Add the spinach. Simmer until the rice is cooked and serve hot.

Rice Pilau *Ann Tatua*

Serves 4

Ingredients

1 cup shelled peas
250 g meat
3 medium carrots
2 large tomatoes
1 tbsp frying fat
3 medium onions
1½ cups rice
½ tsp each black pepper, paprika, cumin seeds, ground cloves, nutmeg, garlic, cinnamon, caraway seeds
1 level tsp salt
1 cup coconut milk

Method

1 Boil the peas. Chop the meat into small pieces and dice the carrots. Blanch and chop the tomatoes and chop the onions.
2 Fry the onions until lightly browned, then add the spices, tomatoes and meat and cook for 10 minutes.
3 Add the carrots, peas and rice. Cover with coconut milk, reduce the heat, cover the pan and cook without stirring until the rice is cooked (about 20 minutes).

Pulses

Pulses or legumes include all types of beans, peas, lentils and groundnuts. They can be used as substitutes for meat and fish because they are cheap and contain a fair amount of vegetable protein. They also contain carbohydrate, calcium, iron, some of the B group of vitamins and roughage. Groundnuts contain fat and vitamins A and B.

All pulses should be soaked and then boiled (soaking reduces the boiling time) to make them soft and digestible. After soaking, they should be put into cold water, boiled and then simmered until tender.

Five cup Irio
Elizabeth W. Kuria

Serves 12

Ingredients

1 cup kidney beans
1 cup black beans
1 cup cowpeas
1 cup pigeon peas
1 cup shelled green maize
4 medium potatoes
salt to taste

Method

1. Soak the kidney beans and black beans overnight. Soak the cowpeas and pigeon peas in warm water for 2–3 hours.
2. Pressure-cook the beans, peas and maize for 40 minutes.
3. Peel the potatoes and add to the mixture. Cook for about 10 minutes.
4. Drain off the water and add salt to taste.
5. Mash until the peas and beans are broken down and mixed with the potatoes. Serve hot.

Muthura
(Irio with Cowpeas and Sorghum)

Elizabeth Karani

Serves 4

Ingredients

½ cup cowpeas
1 cup sorghum grains
5 medium potatoes
2 green bananas (optional)
salt to taste

Method

1. Soak the cowpeas and sorghum overnight. Alternatively, cook them in a pressure cooker for 10 minutes or boil them in a saucepan for 30–40 minutes.
2. Peel the potatoes and bananas.
3. When the cowpeas and sorghum are soft, strain off the excess water, leaving enough to cook the potatoes and bananas.
4. Add the potatoes, bananas and salt to taste and cover.
5. When the potatoes are soft, strain off the water. Heat to dry off any excess water. Mash and serve hot.

Fried Irio

Pauline Mwangi

Serves 2

Ingredients

½ cup dried kidney beans
1 cup green maize
3 medium potatoes
¾ cup pumpkin
1 green banana
1 bunch pumpkin leaves
1 medium onion
¼ cup fat
salt to taste

Method

1. Soak the beans overnight. Boil the beans with the maize for 30–40 minutes or pressure-cook for about 10 minutes.
2. Prepare and chop all the vegetables except the potatoes.
3. Put the fat in a warm saucepan, add the onions and sauté them until soft.
4. Add potatoes, pumpkin and banana and fry, stirring continuously, for about 5 minutes.

5 Add a cup of water or a mixture of tap water and that from the cooked maize and beans.
6 Add salt to taste, then add the pumpkin leaves.
7 Add the maize and beans and cook for 15 minutes. Mash well.

Variations: 1 An extra banana may be added to give a firmer consistency.
2 A chopped tomato may be added after the onions.

Tente
Lilian Muthuuri

Serves 6

Ingredients

1 cup dry pigeon peas (njugu)
1 cup green maize
10 large potatoes
4 bunches pumpkin leaves (optional)
salt to taste

Method

1 Soak the peas in water for a minimum of 3 hours or overnight if possible.
2 Pressure-cook the peas and maize for about 30 minutes or until soft.
3 Peel and chop the potatoes and wash and chop the pumpkin leaves. Put them in a saucepan, pour on enough boiling water to cover them and add salt. Cover and simmer until soft.
4 Drain off the water. Add the maize and peas and return the pan to a very low heat to evaporate all the water.
5 Mash well with a mwiko (wooden spoon). The tente should be quite firm. Serve on a hot dish with meat stew.

Irio with Pigeon Peas
Zipporah Kabugu

Serves 4

Ingredients

1 cup pigeon peas (njugu)
1 cup green maize
2 small potatoes
2 green bananas
1 small onion
1 tbsp fat
1 tsp salt

Method

1. Soak the pigeon peas in warm water for 4 hours.
2. Boil the maize and pigeon peas until tender.
3. Peel the potatoes and bananas and add to the maize and pigeon peas.
4. Add salt and simmer until soft.
5. Drain off excess water and mash.
6. Slice the onion and fry in fat until golden brown.
7. Add the mashed irio and mix well. Serve hot.

Irio with Green Peas

Leah Githinji

Serves 4

Ingredients

1 cup maize
1 cup shelled green peas
7 medium potatoes
salt to taste

Method

1. Boil the maize and peas until soft.
2. Peel the potatoes and cook separately until soft.
3. Mix together the maize, peas and potatoes, mash thoroughly and add salt to taste. Serve with meat stew.

Variations: 1 Fry the irio with onions.
2 Add margarine at stage 3.

Irio with Pumpkin Leaves

Grace Karingithi

Serves 2

Ingredients

$\frac{1}{2}$ cup dried kidney beans
1 cup green maize
1 medium potato
1 cup chopped pumpkin leaves
$\frac{1}{4}$ tsp salt

Method

1. Soak the beans in warm water for 5 hours.
2. Boil the maize in a pressure cooker for 15 minutes.
3. Add the beans to the maize and cook until they are soft.

4 Drain off the water, keeping two cupfuls of it.
5 Put the peeled potato, pumpkin leaves, maize and beans and the two cups of drained water into a saucepan.
6 Boil the mixture until the potatoes are soft and the water has been absorbed.
7 Add salt and mash the mixture. Serve hot with meat stew.

Coconut Beans
Magdalen Musila

Serves 2

Ingredients

1 cup kidney beans
1 tbsp shortening
1 onion
$\frac{1}{2}$ tsp curry powder
$\frac{1}{2}$ tsp dhania
pinch of salt
1 tbsp beef roiko mchuzi mix
1 tsp sugar
2 cups coconut milk

Method

1 Wash the beans and boil them until soft.
2 Fry the chopped onion in the shortening until brown. Add the curry powder, salt, roiko, dhania, tomato, sugar, beans and coconut milk and bring to the boil. Serve hot.

Note: this is a very popular dish among people from the Coast.

Bufuke
Maria Onyango

Serves 2

Ingredients

$\frac{1}{4}$ cup dried kidney beans
$\frac{1}{4}$ cup cowpeas
$\frac{1}{2}$ cup groundnuts
$8\frac{1}{3}$ cups water or stock
3 medium sweet potatoes
1 tsp salt
1 tbsp margarine

Method

1. Soak the beans, peas and groundnuts overnight or for 2–3 hours. Boil in the water or stock until half-cooked.
2. Peel and dice the sweet potatoes. Add them to the pulses and nuts and cook until soft.
3. Add salt and continue to cook to evaporate all the moisture (about 30–40 minutes).
4. Mash with the margarine. Serve hot or cold with stew.

Mukenye
Janet Ngoci

Serves 4

Ingredients

1 cup dry beans
1½ cups green maize
2 medium sweet potatoes
pinch of salt

Method

1. Soak the beans in cold water for 4 hours.
2. Pressure-cook the beans for 30 minutes.
3. Add the maize and cook for 8 minutes.
4. Add the sweet potatoes and cook for 4–5 minutes.
5. Drain off any excess water, add salt and mash. Serve hot with a sauce or vegetable stew.

Muthura Mix
Sylvia Murungi

Serves 5–6

Ingredients

1 cup dry pigeon peas
1 cup crushed sorghum
1 tsp salt
1 medium onion
1 tbsp fat

Method

1. Soak the peas and sorghum for 4 hours.
2. Pressure-cook for 25 minutes. Drain off excess water.
3. Chop the onion and sauté in fat until light brown.
4. Add peas, sorghum and salt to the onion and stir well. Serve immediately.

Mataha ma Kibaki

Dorcas Kimuthia

Serves 4

Ingredients

1 cup dry kidney beans
1 cup green maize
½ bunch greens (kibaki)
2 large potatoes
½ tsp salt
¼ tsp pepper
1 tbsp margarine
tomato to garnish

Method

1. Soak the beans overnight. Boil 1½ cups water, add the beans and simmer for 1 hour.
2. Add the maize and boil for 15 minutes.
3. Chop the greens and peel the potatoes and add them to the maize and bean mixture. Cook for 20 minutes.
4. Drain off the water, add salt, pepper and margarine and mash well.
5. Serve hot, garnished with sliced tomato, with meat stew.

Fried Mataha

Mary Kamami

Serves 2

Ingredients

1 cup rose cocoa beans
2 cups green maize
2 green bananas
2 medium potatoes
1 bunch pumpkin leaves
1½ tsps salt
1 medium onion
2 tsps fat
¼ tsp curry powder

Method

1. Soak the beans overnight.
2. Cook the maize and beans for 30 minutes in a pressure cooker or for 1¼ hours in a saucepan.
3. Peel the bananas and potatoes and wash and chop the pumpkin leaves.
4. Add the potatoes, bananas and pumpkin leaves to the cooked beans and maize. Cook for a further 10–15 minutes or until soft.

5 Drain and mash. Add the salt.
6 Slice the onion, fry gently until tender and add the curry powder. Then add the mashed mataha, mix well and cook for 1-2 minutes. Serve hot.

Muree Mash
(Cowpea Mash)

Gertrude Irere

Serves 2

Ingredients

1½ cups cowpeas
2 medium potatoes
3 raw bananas
3 tbsps fat
1 medium onion
1½ cups water
1 tsp salt
parsley to garnish

Method

1 Break up the dry cowpeas using a grinding stone, blow off the husks and wash well. Boil until just tender, drain off all the water and leave covered.
2 Peel and slice the potatoes. Peel the bananas and leave in salted water to prevent discoloration.
3 Fry the onions, then sauté the potatoes and bananas in the fat and add the water. Bring to the boil and add the cowpeas.
4 Cook until tender, add salt and drain off the liquid. Mash well and serve, sprinkled with parsley, with meat curry.

Traditional Isyo

Florine Mbithi

Serves 4

Ingredients

1½ cups dry beans
1 cup dry maize
2 bunches terere (or any green leafy vegetable)
1 medium cassava
salt to taste

Method

1 Soak the beans in hot water for 3-4 hours or overnight.
2 Wash the maize and boil for 45 minutes, then add the beans and boil for a further 45 minutes.
3 Add the prepared terere, salt and chopped cassava. When the cassava is tender, drain off the water and mash.

Isyo sya Nthooko
Josephine Kavilu

Serves 2

Ingredients

- ¾ cup dried cowpeas
- ¾ cup green maize
- 1 cup chopped cowpea leaves (kunde)
- 1 tsp salt
- 1 medium onion
- 1½ tbsps fat for frying
- 1 tsp curry powder
- 1 tbsp margarine

Method

1. Soak the cowpeas for 4-6 hours in warm water before cooking.
2. Cook the maize in 6-8 cups water in a pressure cooker for 15-20 minutes.
3. Wash the cowpeas and add to the maize. Boil for 15-20 minutes.
4. Add the chopped kunde leaves. Cook until tender. (If the kunde leaves are not young, they should be cooked separately until tender before adding them to the maize and peas.)
5. Strain off excess water and add the salt.
6. Fry the chopped onion in fat, add the curry powder and margarine. Add to the maize, cowpea and kunde mixture.
7. Simmer for 5 minutes. Serve hot.

Sweet Njahe
Charity Mwangi

Serves 4

Ingredients

- 1½ cups njahe (black beans)
- ½ cup maize
- 3 green bananas
- 2 ripe bananas
- 1 tsp sugar
- salt to taste

Method

1. Soak the beans overnight.
2. Boil the beans and maize in a pressure cooker for 35-40 minutes or until the beans are soft.
3. Peel the green bananas and add to the maize and beans.
4. Boil for a further 20 minutes, or until the bananas are soft.
5. Mash well; peel the ripe bananas and add to the mixture with the salt and sugar. Continue mashing until thoroughly mixed. Serve hot on its own or with vegetables.

Njahe cia Muciairi

Rachel Burugu

Serves 2

Ingredients

2 cups njahe (black beans)
2 cups green maize
4 green bananas
½ cup sugar
1 tsp salt

Method

1. Pressure-cook the beans for 30 minutes. Retain the water.
2. Pressure-cook the maize for 10 minutes.
3. Peel the bananas and boil for 20 minutes.
4. Mix the cooked beans, maize and bananas in a saucepan and cook for 10 minutes in the water used for cooking the beans.
5. Add the sugar and salt and mash thoroughly. Serve hot with meat stew.

Note: this dish is suitable for a lactating mother.

Nyoyo

Rose Osungu

Serves 2

Ingredients

¼ cup cowpeas
¼ cup groundnuts
¼ cup greengrams
1 cup green maize
1 tbsp kimbo
1 medium onion
1 medium tomato
¼ tsp mixed spice
¼ tsp curry powder

Method

1. Soak the cowpeas overnight.
2. Boil the groundnuts, greengrams, green maize and cowpeas in a pressure cooker for 10–15 minutes or in a saucepan for 40 minutes until soft.
3. Melt the kimbo in a heavy pan, add the chopped onion and fry until soft. Add the cowpea mixture to the onions, with the chopped tomato and seasonings.
4. Heat for a further 5 minutes or until there is no excess water. Serve hot.

Bean and Sweet Potato Mix

Miriam Sei

Serves 4

Ingredients

1 cup beans
1 cup green maize
1 medium green pepper
1 medium onion
1 medium tomato
2 medium carrots
2 large sweet potatoes
1 tbsp cooking oil
1 tsp curry powder
1 heaped tsp margarine
1 bunch spinach
salt to taste

Method

1. Boil the pre-soaked beans and green maize until tender.
2. Wash and slice the green pepper, onion and tomato. Dice the carrots and sweet potatoes.
3. Fry the onions and green pepper in a deep saucepan until tender. Add the curry powder, chopped tomato and carrots and cook for 5 minutes.
4. Add the sweet potatoes and mix well. Add the cooked bean-maize mixture and enough of the water it was cooked in to cover the mixture. Add the margarine.
5. Simmer for 15 minutes, or until the potatoes are tender. Sauté the spinach, add salt and serve on top of the mixture.

Kidney Bean Mash

Rahab Kamau

Serves 4

Ingredients

1 cup white kidney beans (nowe)
4 medium potatoes
pinch of salt
1 tbsp chopped onion
1 tbsp chopped dhania
1 tsp margarine
sliced tomatoes to garnish

Method

1. Boil the pre-soaked beans for 30 minutes or until they are soft.
2. Peel the potatoes, add them to the beans and boil until the potatoes are cooked.
3. Drain off the water, add salt and mash thoroughly.
4. Sauté the onion and dhania and add to the bean mash.
5. Serve garnished with sliced tomatoes.

Variations: 1 Add $\frac{1}{2}$ cup cooked green maize at stage 3.
2 The mash may be piped decoratively and grilled before serving.

Mulee
(Traditional Meru)

Janet Ngoci

Serves 2

Ingredients

1 cup dried cowpeas
3 green bananas
4 medium potatoes
1 tsp salt
1 tbsp margarine

Method

1. Break the cowpeas up using a grinding stone. (The stone should be passed over the cowpeas hard enough to break them into pieces but not to grind them into flour.) Boil for 10 minutes.
2. Peel and chop the bananas and potatoes. Add them to the cowpeas and continue cooking until tender.
3. Drain off the excess water. Add the salt and margarine and mash. Serve hot with a sauce and green vegetables.

Improved Mulee

Janet Ngoci

Serves 3

Ingredients

1 cup dried cowpeas, ground
250 g meat
1 onion
1 tbsp margarine
3 green bananas
4 medium potatoes

½ tsp curry powder
2 tomatoes
1 tsp salt
parsley to garnish

Method

1. Boil the cowpeas in 4 cups of water for 10–15 minutes.
2. Chop the meat and onions and fry together until the meat is brown. Add the chopped tomatoes and curry powder and 3 cups of water. Cook on a low heat for 15–20 minutes.
3. Peel and chop the bananas and potatoes. Add to the meat and continue cooking for 10–12 minutes or until tender.
4. Add the cowpeas. Reduce the heat and simmer for 5 minutes.
5. Remove from the heat and garnish with parsley. Serve hot with green vegetables.

Creamed Njenga   *Margaret Ngau*

Serves 2

Ingredients

1 cup njenga
4 cups water
1½ tsp salt
1 tbsp margarine
2 heaped tbsps maize flour
1 ripe banana and parsley to garnish

Method

1. Sieve the njenga and wash it in several changes of water.
2. Put 4 cups of water in a saucepan, add the salt and bring to the boil. Add the njenga and half the margarine. Bring to the boil, then simmer for 10–15 minutes, stirring occasionally. Cook until tender but not dry, adding margarine as required.
3. Thicken by adding maize flour a little at a time: the njenga should not be too stiff. Garnish with banana and parsley and serve with stew, ndengu (greengrams) sauce, cowpea sauce, sour milk or vegetables.

Note: njenga is the ramains of sifted, crudely ground maize. It can be used as a rice substitute.

Tubers

The most common tubers in Kenya are cassava, sweet potatoes, yams and English potatoes.

A wide variety of dishes may be prepared from cassava flour or whole cassavas. They are very high in carbohydrate, hence a good source of energy, but contain a negligible amount of protein. They have traditionally been used as a famine crop. The cassava tuber contains a toxic substance (hydrocyanic acid) which is present beneath the cassava peel. Processing, which includes peeling, washing, slicing, grating and cooking, renders cassava fit for human consumption. It is not advisable to eat it raw.

Sweet potatoes are popular in Kenya. They may be eaten as part of a main meal, in stews and soups or as a snack. They contain some protein, highly digestible carbohydrate and some vitamins B and C.

A wide variety of yams is eaten in Kenya. They are usually peeled, sliced, steamed and served with soups or sauces, and sometimes peeled, sliced and fried as a snack. Yams are rich in carbohydrate and contain traces of mineral salts, vitamins and protein just beneath the peel.

Green bananas come under the category of fruits, but they are served as a starchy food. Green bananas may be used as yam or potato substitutes in pounded or mashed dishes. For good results, bananas should be mixed with other starchy foods such as yams. They provide carbohydrate and small quantities of vitamins B_1 and B_2.

Glazed Sweet Potatoes

Rachel Burugu

Serves 2

Ingredients

4 medium-sized sweet potatoes
$\frac{1}{2}$ cup fat

Method

1. Boil the potatoes until they are nearly soft. Remove the outer skin and cut into slices 1 to 2 cm thick.
2. Shallow-fry the slices or brush them with melted fat and grill until golden brown on both sides.
3. Serve hot with meat stew and sliced tomatoes for lunch or supper, or on their own for breakfast.

Sweet Potatoes Baked in Margarine
Grace Karing'ithi

Serves 4

Ingredients

2 medium sweet potatoes
$\frac{1}{4}$ cup melted margarine
2 tbsps sugar
$\frac{1}{2}$ tsp salt
$\frac{1}{2}$ tsp pepper

Method

1. Peel and boil the sweet potatoes until they are soft. Allow them to cool and slice them lengthwise.
2. Pre-heat the oven to 400 °F (200 °C, gas mark 6).
3. Mix together the melted margarine, sugar, salt and pepper, and dip the slices of sweet potato into the mixture.
4. Put the slices on a baking sheet and bake them on the top shelf of the oven until brown (about 10–12 minutes). Serve hot with vegetable stew or greengram sauce.

Grilled Sweet Potatoes
Eunice Kirimi

Serves 2

Ingredients

2 large sweet potatoes
2 medium onions
2 medium tomatoes
1 tbsp margarine
pinch of pepper

Method

1. Wash and boil the sweet potatoes. Peel and slice the onions and tomatoes.
2. When the potatoes are cooked, peel them, brush with melted margarine, sprinkle with pepper and grill until golden brown on all sides. Grill the onions and tomatoes with the potatoes. Serve hot.

Mukenye Sweet Potato Rolls

Magdalen Juma

Serves 4

Ingredients

2 sweet potatoes
1 cup greengrams
1 cup green peas
¼ tsp black pepper
salt and pepper to taste
1 tsp baking flour
1 egg, beaten
1 cup fine breadcrumbs
oil for deep frying

Method

1. Wash peel, chop and boil the potatoes.
2. Boil the greengrams and peas separately.
3. When ready, mash all three with pepper, salt and flour. Shape into circular cakes.
4. Coat first in beaten egg and then in breadcrumbs, and deep-fry until golden brown. Serve hot or cold.

Variation: they can be cooked in the oven for 20 minutes at 425 °F (220 °C, gas mark 7), or under the grill.

Sweet Potato and Greengram Mash

Leah Kibiego

Serves 2

Ingredients

1 cup greengrams
2 medium-sized sweet potatoes
½ tsp salt

Method

1. Clean the greengrams and boil them until almost soft.
2. Peel the sweet potatoes and cut them into small cubes. Add to the greengrams and cook until soft.
3. Mash and serve hot with any sauce or fresh or sour milk.

Sweet Potato Balls
Jane Kuria

Serves 4

Ingredients

3 medium-sized sweet potatoes
2 eggs, beaten
1 medium onion
½ cup breadcrumbs
salt and pepper to taste
oil for deep frying

Method

1. Peel and boil the sweet potatoes.
2. Chop the onion, fry in 1 tbsp oil and add salt and pepper.
3. When the sweet potatoes are soft, mash them and add to the onion.
4. Roll the sweet potatoes into balls. Coat each ball with the egg, then roll it in the breadcrumbs.
5. Deep-fry until golden brown. Serve hot.

Mashed Sweet Potatoes
Dorcas Male

Serves 2

Ingredients

2 small sweet potatoes
1 tbsp margarine
2 tbsps chopped onion
2 tbsps flour
½ cup milk

Method

1. Peel and chop the sweet potatoes and boil them until soft. Drain and mash.
2. Melt margarine, add onion and cook for about 2 minutes. Add flour and cook for about one minute without browning.
3. Add mashed sweet potatoes. Add milk and mix thoroughly.
4. Press into a greased mould and bake in a moderate oven (180 °C, 350 °F, gas mark 4) until brown.

Enriched Mashed Sweet Potatoes

Alice Wafula

Serves 2

Ingredients

2 sweet potatoes
¼ tsp salt
1 tbsp margarine or butter
1 tbsp milk

Method

1 Peel the sweet potatoes and cut them into large pieces (red potatoes will give the best results).
2 Boil in salted water in a covered saucepan until soft.
3 Add the margarine or butter and milk and mash until soft.
4 Serve hot with meat stew and green vegetables.

Sweet Potatoes with Meat

Julie Mugo

Serves 2

Ingredients

250 g beef
1 onion
2 tbsps margarine
1 clove garlic
½ tsp curry powder
3 tomatoes
salt to taste
2 sweet potatoes

Method

1 Prepare and chop the meat. Chop the onion finely and fry in 1 tbsp of margarine.
2 Add the finely chopped garlic, curry powder, blanched and chopped tomatoes and meat. Season with salt and cook until tender.
3 Wash and peel the sweet potatoes, boil until tender but still firm, and chop them into large cubes.
4 Heat the remaining margarine and fry the sweet potatoes until golden brown. Serve hot with the beef stew.

Mashed Potatoes

Leah Githinji

Serves 3

Ingredients

6 medium potatoes
1 tsp margarine
¼ tsp cinnamon
salt and pepper to taste
2 tbsps milk

Method

1. Peel the potatoes and boil until soft.
2. Drain off the water and add the margarine, cinnamon, salt and pepper to the potatoes.
3. Mash, adding milk, until smooth. Serve with meat stew and another vegetable.

Arrowroot Balls

Elizabeth W. Kuria

Serves 6

Ingredients

3 medium arrowroots
1 medium onion
1 egg
4 tbsps breadcrumbs
fat for deep-frying

Method

1. Peel the arrowroots and boil until soft.
2. Chop and fry the onions in 3 tbsps of the fat.
3. Mash the arrowroots and add them to the fried onion.
4. Roll the fried arrowroots into small balls. Coat with the beaten egg, then roll them in the breadcrumbs.
5. Heat the fat and deep-fry for 3 minutes or until golden brown. Serve hot.

Nduma Casserole

Jennifer Kaniaru

Serves 6

Ingredients

3 nduma (arrowroots)
3 medium potatoes
1 onion
1 tbsp fat
1 tsp salt
1 tsp butter

Method

1. Peel and dice the nduma and potatoes, cutting the nduma into larger pieces than the potatoes. Keep them separately and cover with water.
2. Chop the onion finely and fry it in a sufuria until light brown.
3. Put the potatoes into the sufuria. Stir gently, then cover for 1 minute. Add enough water to cover them, replace the lid and let them simmer.
4. Using a sharp knife, check to see whether the potatoes are almost ready. Add the nduma and more water if necessary to cover the vegetables. Simmer gently until cooked (about 20 minutes).
5. Put into a casserole dish, add butter and cook in the oven for 15 minutes at 350 °F (180 °C, gas mark 4). Serve hot.

Nduma Casserole with Cheese

Rahab Kamau

Serves 2

Ingredients

For the casserole:
1 small arrowroot
2 rashers of bacon

For the cheese sauce:
1 tbsp margarine
2 tbsps flour
2 cups milk
$\frac{1}{4}$ cup grated Cheddar cheese
salt and pepper to taste

Method

1. Wash, dice and boil the arrowroot until tender.
2. Cook the bacon on a low heat.

3 Arrange the arrowroot, bacon and cheese alternately in a casserole with the arrowroot at the bottom and the cheese at the top.
4 Prepare the cheese sauce and pour over the dish.
5 Cook in a moderately hot oven (370 °F, 190 °C, gas mark 5) for 5 minutes. Serve immediately.

Variation: add cooked carrots and peas.

Cheese Sauce:
1 Melt the margarine in a saucepan.
2 Remove the pan from the heat and stir in the flour until it forms a smooth paste.
3 Add milk, a little at a time, stirring continuously.
4 Heat gradually until it boils, stirring all the time.
5 Add cheese, salt and pepper and simmer for 2–3 minutes.

Other vegetables

Vegetables fall into three main categories:
1 Green leafy vegetables, e.g. sukumawiki, cowpea leaves.
2 Fruit vegetables, e.g. tomatoes, peppers.
3 Roots and tubers, e.g. carrots, onions, potatoes.

Nutrients are found in all parts of vegetables – leaves, stems, fruits and roots. They can be valuable sources of vitamin C (tomatoes), vitamin A (carrots and greens), minerals (potatoes and carrots) and some proteins (beans).

They are valuable sources of roughage, which promotes digestion and helps to neutralise acid substances produced during digestion. Vegetables lose their nutritive value if stored for a long time.

In order to get the best nutritive value from vegetables, choose those that are fresh and firm. Vegetables should be washed in plenty of water before peeling or cutting. Avoid soaking them in water; this leads to loss in nutritive value.

When possible, vegetables should be cooked in their skins. This prevents the nutrients just beneath the skin escaping into the water. Tomatoes should be blanched before cooking. Vegetables should be cooked in small amounts of water.

Mixed Green Vegetables

Violet Mugalavai

Serves 4

Ingredients

4   okra
$2\frac{1}{2}$ cups water
2   bunches kunde
2   bunches terere
1   tbsp fat
1   medium onion
2   medium tomatoes
salt to taste

Method

1 Wash and chop the okra, kunde and terere.
2 Boil the water and add the kunde, terere and okra. Cook for 10 minutes.
3 Melt the fat and fry the onion until tender. Add the chopped tomatoes, green vegetables and salt and fry for 5 minutes. Serve sizzling hot.

Steamed Vegetables

Margaret Racho

Serves 6

Ingredients

3 bunches deg akeyo
1 bunch omboga
$\frac{1}{3}$ cup water
$\frac{1}{2}$ onion, diced
4 tbsps ghee
1 medium tomato
$\frac{1}{4}$ cup milk
$\frac{1}{4}$ tsp salt

Method

1. Remove and discard the stalks of the green vegetables, wash and shred the leaves.
2. Pressure-cook them with $\frac{1}{3}$ cup water for 10 minutes, then reduce the heat and cook for a further 5 minutes or until the strong odour has gone.
3. Mash the vegetables.
4. Fry the onion in the ghee until golden brown, add the blanched and chopped tomato, then the green vegetables.
5. Add the milk and salt and simmer for 5 minutes. Serve hot with ugali or sweet potatoes.

Note: deg akeyo is a bitter vegetable. Some traditional methods of preparation involve cooking it in several changes of water.

Fried Sukumawiki

Joyce Onyango

Serves 2

Ingredients

10 young sukumawiki leaves
$\frac{1}{2}$ tbsp cooking fat
$\frac{1}{2}$ tsp salt

Method

1. Wash the sukumawiki thoroughly in cold running water. Remove the stalks and shred the leaves.
2. Heat the fat and add the sukumawiki, stirring gently.
3. Add the salt and cook on a low heat, stirring constantly, for 5 minutes. Serve hot with ugali.

Variation: add 1 tsp peanut butter just before removing from the heat.

Other vegetables 117

Sukumawiki in Groundnut Sauce
Jane Odipo

Serves 2

Ingredients

1 large onion
1 bunch sukumawiki
1½ tbsps cooking fat
salt to taste
1 tbsp groundnut paste

Method

1 Clean and chop the onion and sukumawiki.
2 Sauté the onion in hot fat, then add the sukumawiki. Fry for 5 minutes (longer if the sukumawiki is tough).
3 Add 1 tbsp water and leave to cook for a further 5 minutes.
4 Add salt to taste, then the groundnut paste and mix thoroughly. Serve with ugali.

Note: any green leafy vegetable can be cooked in this way.

Fried Egg Spinach
Asenath Sigot

Serves 2

Ingredients

1 bunch spinach
1 medium onion
1 tbsp cooking fat
2 eggs
½ tsp salt
¼ tsp pepper

Method

1 Wash and chop the spinach.
2 Cook in a covered saucepan for 1 minute, add the salt and stir. Cook for a further 2 minutes and remove from the heat.
3 Fry the chopped onion in the fat until golden brown. Beat the eggs, pour them over the cooked onion and let them cook on a low heat for 2 minutes.
4 Spread the cooked spinach over the eggs, add salt and pepper, then stir to mix. Serve immediately.

Likhubi
(Cowpea Leaves)

Mary Munyole

Serves 16

Ingredients

8 bunches young cowpea leaves
1 medium onion
1 tbsp fat
$\frac{1}{2}$ cup water
4 tbsps peanut butter
$\frac{1}{2}$ cup fresh boiled milk
salt to taste

Method

1. Remove and discard the cowpea leaf stems. Wash, drain and chop the leaves.
2. Finely chop the onion and fry at medium temperature in the fat until light brown.
3. Add the cowpea leaves to the onion, a little at a time.
4. Add half the water, cover and cook for 5 minutes.
5. Turn the vegetables, cover and cook for a further 5 minutes.
6. Add the remaining water, turn, cover and cook for a further 5 minutes.
7. Stir in the peanut butter until smooth.
8. Remove from the heat, add the milk and mix thoroughly. Add salt and serve hot with ugali and meat.

Cowpea Leaves in Mushroom Sauce

Catherine Ochoki

Serves 8

Ingredients

1 cup chopped mushrooms
$\frac{1}{2}$ cup milk
salt to taste
4 bunches cowpea leaves
$\frac{1}{2}$ cup water
$\frac{1}{2}$ tsp bicarbonate of soda or traditional salt
$\frac{1}{2}$ tbsp fat (optional)

Method

1. Cook the chopped mushrooms until tender. Stir in the milk and salt and remove from the heat.
2. Wash the cowpea leaves.
3. Boil the water and add the bicarbonate of soda.
4. Add the cowpea leaves and simmer for about 5 minutes, stirring constantly.

5 Drain the cowpea leaves and add them to the mushrooms.
6 Add fat if desired and simmer for 5 minutes. Remove from the heat and serve hot with ugali or mashed bananas.

Note: traditional salt is ash made from dried and burnt green leafy vegetables.

Cowpea Leaves in Groundnut Sauce *Bertha Ochako*

Serves 2

Ingredients

2/3 cup water
1/2 tsp bicarbonate of soda
1 bunch cowpea leaves
2 tsps ghee
3/4 cup groundnuts, pounded
1 cup milk
salt to taste

Method

1 Boil the water and bicarbonate of soda.
2 Wash and chop the cowpea leaves and add them to the boiling water. Boil until dry.
3 Melt the ghee in a saucepan, add the groundnuts and milk and stir vigorously.
4 Add the cowpea leaves, season and simmer gently for 40 minutes, stirring occasionally. Season to taste and serve hot with ugali.

Apoth in Groundnut Sauce *Damari Otieno*

Serves 2

Ingredients

1/2 bunch cowpea leaves (kunde)
1 small bunch apoth
2 1/2 cups water
1 1/2 tsp salt
1/2 tsp bicarbonate of soda
1 cup pounded groundnuts

Method

1 Wash and shred the kunde and apoth. Boil 2 cups of salted water with soda.
2 Add the apoth to the boiling water and cook for 1 minute.

3 Add the kunde and boil for 15 minutes, stirring once.
4 Make a paste with the groundnuts and $\frac{1}{2}$ cup warm water. Add to the kunde and simmer for 5 minutes. Serve with ugali.

Murere Mix

Mary Munyole

Serves 2

Ingredients

1 bunch murere
$\frac{1}{2}$ cup water
pinch of salt
$\frac{1}{4}$ tsp bicarbonate of soda
1 small onion
1 tsp margarine
$\frac{1}{2}$ cup milk

Method

1 Remove and discard the stems of murere and wash the leaves under cold running water.
2 Boil the water with the salt and bicarbonate of soda. Add the murere and simmer for 10 minutes.
3 Finely chop the onion and fry in the margarine until tender.
4 Add the onion and milk to the murere and boil for 2–3 minutes. Serve hot with cooked mashed bananas, ugali, sweet potatoes or beans.

Terere Dish

Pauline Mudek

Serves 4

Ingredients

2 bunches terere
$\frac{1}{2}$ cup milk
$\frac{1}{2}$ cup water
salt to taste

Method

1 Wash the terere leaves under running water.
2 Boil $\frac{1}{2}$ cup water and add the terere a handful at a time. Simmer for 20 minutes.
3 Drain the terere and simmer in the milk for 5 minutes. Add salt to taste. Serve immediately with ugali.

Nutty Pumpkin

Anne Masenge

Serves 3

Ingredients

¼ pumpkin
2 bunches fresh pumpkin leaves
½ cup groundnuts, pounded
¼ tsp salt
pinch of curry powder (optional)

Method

1. Wash the pumpkin and pumpkin leaves.
2. Remove the pumpkin skin and seeds and chop the flesh.
3. Remove and discard the stems of the pumpkin leaves and chop the leaves. Put the leaves and the pumpkin in a saucepan, add water and cook until the pumpkin is soft and only a little water remains in the pan. Add the pounded groundnuts, salt and curry powder and mash to a smooth consistency. Serve hot or cold.

Fried Pumpkin Leaves

Janet Olubayo

Serves 4

Ingredients

2 bunches pumpkin leaves
1 medium onion
2 medium tomatoes
½ cup water
¼ tsp bicarbonate of soda
1 tbsp fat
½ tsp salt
1 tbsp peanut butter

Method

1. Prepare and shred the pumpkin leaves and chop the onion. Blanch, skin and chop the tomatoes.
2. Boil the pumpkin leaves in water and bicarbonate of soda for 3 minutes.
3. Fry the onion in fat until brown.
4. Add the tomatoes and salt and continue cooking for about 3 minutes.
5. Add the pumpkin leaves and simmer for 3 minutes. Add the peanut butter and simmer briefly. Serve hot with ugali.

Fried White Cabbage

Leunita Muruli

Serves 4

Ingredients

1 medium-sized white cabbage
1 medium onion
2 medium tomatoes
2 tbsps cooking fat
$\frac{1}{2}$ cup boiled milk
salt to taste

Method

1. Wash and shred the cabbage, chop the onion and slice the tomatoes.
2. Fry the onion in the fat at medium temperature until brown.
3. Add the cabbage, cover and cook for 3 minutes.
4. Turn the cabbage, add the sliced tomatoes and cover.
5. Cook for 3 minutes and remove from the heat.
6. Add the milk and mix well. Season to taste and serve hot with rice, ugali or boiled potatoes.

Crunchy Cooked Cabbage

Ruth Oniang'o

Serves 3

Ingredients

1 small cabbage
1 onion
$1\frac{1}{2}$ tbsps cooking fat
2 tbsps water
salt to taste

Method

1. Wash and finely shred the cabbage and chop the onion.
2. Sauté the onion in the hot fat until tender.
3. Add the cabbage and water and cook for 6 minutes, stirring once. Season and serve with meat and ugali.

Sautéed Green Beans

Linda Ethangatta

Serves 6

Ingredients

2 rashers bacon, chopped
3 cups green beans, chopped
½ medium onion, sliced
salt to taste

Method

1. Fry the bacon until brown. Remove from the pan and drain on absorbent paper.
2. Add onion to the fat and fry until tender. Add beans and salt to taste.
3. Cook until the beans are tender but still pale green. Garnish with the bacon.

Mashed Bananas

Elizabeth Nafula Kuria

Serves 6

Ingredients

12 green bananas
1 cup water
pinch of salt
¼ tsp curry powder
2 tbsps margarine
2 tbsps warm milk

Method

1. Peel and chop the bananas.
2. Bring the salted water to the boil. Add the bananas and curry powder and simmer for 30 minutes or until soft.
3. Add margarine and warm milk and mash thoroughly. Serve hot.

Bananas in Coconut

Jane Odipo

Serves 2

Ingredients

5 cooking bananas
1 large onion, chopped
1 tomato, chopped

1½ tbsp cooking fat
1 cup coconut milk
salt and pepper to taste

Method

1 Peel and slice the bananas.
2 Sauté the onion until tender, then add the tomato and sliced bananas; mix well and add the coconut milk.
3 Season and cook until the bananas are tender.

Ngunza Matu (1)
(Embu)

Charity Njiru

Serves 2

Ingredients

¼ cup njahe
½ ripe pawpaw
1 bunch pumpkin leaves
2 tbsps margarine
1 tsp salt
½ cup maize flour
2 cups water

Method

1 Presoak the njahe and boil until tender.
2 Prepare and chop the pawpaw. Add to the njahe and cook for 20 minutes.
3 Chop the pumpkin leaves and cook until tender in melted margarine. Add the salt.
4 Add the flour to make a thick paste. Cook for 20 minutes. Serve hot.

Ngunza Matu (2)
(Kamba)

Theresia Kinai

Serves 2

Ingredients

½ cup chopped kivwea (Kamba) or terere (Kikuyu)
1 tbsp cooking fat
1 medium onion
½ tbsp chopped dhania
¾ tsp salt
¼ tsp pepper
2 cups water
½ cups maize meal

Method

1 Wash the terere thoroughly. Remove and discard the stems and chop the leaves very finely.
2 Melt the fat, add the finely chopped onion and fry until soft but not brown.
3 Add the terere and dhania and mix well. Cook for about 15 minutes or until the leaves are soft.
4 Add the water, salt and pepper and bring to the boil.
5 When the water boils, gradually add the sifted maize meal, stirring continuously. Cook over a medium heat for 20-25 minutes. Serve very hot.

Sauces

A sauce is a liquid used to add flavour and/or extra nourishment to some food. There are three types: pouring sauces, coating sauces and binding sauces or panadas. A pouring sauce is used as an accompaniment and should be fluid enough to be poured from a container. A coating sauce is thicker; it should be smooth enough to be poured over food but of a thick enough consistency not to slide off it. A binding sauce is too thick to pour and is used to bind ingredients together.

The ingredients of a basic white sauce are flour, fat and a liquid, usually milk and water or milk and stock. Equal quantities of flour and fat are used, and the quantity of liquid will depend on the consistency required. There are two ways of making a basic sauce: by blending and by making a roux.

Blending

1. Mix a little of the liquid with cornflour.
2. Bring the rest of the liquid to the boil and add the seasoning, flavouring and fat.
3. Pour the boiling liquid onto the cornflour, stirring all the time.
4. Pour the sauce back into the pan and heat slowly, stirring all the time, for a few minutes.

The roux method

1. Melt the fat gently, stir the plain flour into it and stir over a low heat for 1 or 2 minutes.
2. Take the pan off the heat and gradually add half the liquid, stirring all the time.
3. Add the rest of the liquid and return to the heat. Bring the sauce slowly to the boil, stirring all the time, and boil gently for 3 or 4 minutes. Add seasoning and flavouring.

Sitayani
(Bean Sauce)

Janet Olubayo

Serves 2

Ingredients

1 cup beans
¼ tsp bicarbonate of soda
1 tbsp cooking fat
½ tsp salt

Method

1. Toast the beans with 4 tbsps of water. Dry them and remove the outer covering by rubbing between the hands.
2. Soak the beans for a few minutes before cooking.
3. Boil the beans until tender. Add the bicarbonate of soda and continue boiling for 15 minutes. A thick sauce should form. Add the salt and fat and allow to simmer. Serve hot with ugali.

Groundnut Sauce

Joy Awori

Serves 6

Ingredients

1½ cups groundnuts, pounded
3 cups water
salt to taste
1 onion
2 tomatoes
1 bunch spinach

Method

1. Mix the pounded groundnuts with the water and cook, stirring constantly, until fat collects on the top (about 20 minutes). If the mixture becomes too thick add a little more water.
2. Add salt, chopped onion, tomatoes, and spinach. Continue cooking until spinach is cooked and the sauce is of a thick consistency. Serve with mashed cooked bananas or sweet potatoes.

Cowpea Leaf Sauce

Lynette Opondo

Serves 1

Ingredients

2 cups water
$\frac{1}{4}$ tsp bicarbonate of soda
$\frac{1}{4}$ cup finely chopped murere
$\frac{3}{4}$ cup finely chopped cowpea leaves
1 tsp fat
$\frac{1}{4}$ tsp salt
$\frac{1}{2}$ cup milk

Method

1 Boil the water, bicarbonate of soda and murere for 2 minutes.
2 Add the cowpea leaves and boil for a further 2 minutes.
3 Melt the fat and add the drained vegetables, salt and milk. Simmer for 5 minutes. Serve hot with ugali.

Cowpea Sauce

Margaret Ngau

Serves 2

Ingredients

$\frac{1}{2}$ cup dried cowpeas
1 medium onion
1 tbsp kimbo
$\frac{1}{4}$ tsp mixed spice
$\frac{1}{4}$ tsp curry powder
125 g chopped beefsteak
1 large tomato
1 tsp salt
grated carrot to garnish

Method

1 Soak the cowpeas for at least 6 hours in warm salted water, then skin them.
2 Finely chop the onion and fry it in kimbo until light brown. Add the curry powder and mixed spice and stir well.
3 Add the meat and cook on a low heat until tender.
4 Blanch and chop the tomato.
5 Pressure-cook the cowpeas for 20–25 minutes. Whisk them to a smooth consistency.
6 Add the tomatoes to the meat and continue cooking until the tomatoes are creamed. Pour the cowpea sauce over the fried meat, add salt and mix well. Serve garnished with grated carrot.

Simsim Sauce

Mary Makokha

Serves 12

Ingredients

3 cups simsim seeds
4 bunches cowpea leaves (kunde)
1 onion
1 tomato
2 tbsps cooking fat
4 cups water
1 tbsp curry powder
1 tsp salt

Method

1. Pick stones and grit from the simsim seeds and toast them. Pound to a fine paste.
2. Wash and finely chop the cowpea leaves.
3. Chop the onion and tomato and sauté with the cowpea leaves for 3–5 minutes.
4. Add the water and curry powder and simmer for 10 minutes.
5. Add the simsim paste and simmer, stirring, for 2 minutes. Add the salt and simmer for 5 minutes.
6. Serve hot with ugali and sour milk, sweet potatoes or cooked bananas.

Tomato Sauce

Ruth Oniang'o

Ingredients

about 18 ripe tomatoes
3 tsps salt
1 tsp paprika
pinch of cayenne pepper
2 cups spiced white wine vinegar
1 tbsp tarragon vinegar
$2\frac{1}{2}$ cups granulated sugar

Method

1. Wash, slice and cook the tomatoes slowly until pulped.
2. Rub through a sieve, add the salt, paprika and cayenne and cook gently until the mixture begins to thicken.
3. Add both kinds of vinegar and continue to cook gently until thick and creamy.
4. Add the sugar and reduce until the sauce does not separate.

Bottling:
1. Fill the warmed bottles to 2 cm from the top.
2. Tie the corks on firmly or use screw caps.
3. Stand the bottles in a deep pan of boiling water and simmer.
4. Then screw the corks or caps tight and store in a cool place.

Approximate yield: $2\frac{1}{2}$–3 700ml bottles.

Syrup

Ruth Oniang'o

Ingredients

$1\frac{1}{2}$ cups sugar
pinch of salt
1 cup boiling water
juice of 1 lemon

Method

1. Mix the sugar with the salt, add the boiling water and lemon juice.
2. Heat until froth forms (about 10 minutes) Allow to cool slightly before using. Serve with pancakes.

Bread

Bread is a mixture of flour and liquid, which is raised by adding yeast. Yeast is a living substance that ferments when sugar and liquid are added to it, producing bubbles of carbon dioxide. When dough is made it should be put in a warm place; the warmth makes the bubbles expand and the dough rise. When the dough is baked, the bubbles expand more and the bread rises further. Yeast 'works' best at 25 °C (80 °F) so when making bread it is best to use slightly warmed ingredients.

Basic Bread
Asenath Sigot

Ingredients

- 1½ tsps yeast
- 1 tbsp sugar
- 1½ cups lukewarm water
- 4 cups wheat flour (white or wholemeal)
- 1 tsp salt
- 1 tbsp oil

Method

1. Put the yeast and sugar in a cup, add ½ cup lukewarm water and leave it to stand for 10 minutes until the yeast is completely dissolved.
2. Sift 2 cups of flour into a bowl, add the salt and yeast and mix well. Add 1 cup lukewarm water and mix well.
3. Knead the mixture until it is smooth and soft. Add more flour while kneading.
4. Sprinkle a little flour into a container and put the dough on it. Spread about 2 tsps oil on the surface of the dough to prevent it drying.
5. Cover the container with a wet towel and put it in a warm place for about 1 hour to let the dough rise.
6. When the dough has risen, knead it again, and add more flour if necessary. Grease a baking tin and put the dough into it. Cover it with a wet towel and leave it in a warm place for about 30 minutes to rise again.
7. Put the tin of dough in a moderate oven (350 °F, 180 °C, gas mark 4-5) and bake for 30-40 minutes, until brown.
8. When the bread has baked, spread a little fat on top of the crust, leave for a couple of minutes and then take it out of the oven.
9. Take the bread out of the tin and let it cool.

Variations:
1. Use equal amounts of white and wholemeal flour.
2. Use milk instead of water.
3. Substitute 2 tbsps honey for the sugar.
4. Add 4 tbsps margarine to the dough.
5. Add 1 cup of raisins to the dough before kneading it for the second time.
6. Bread can be made into different shapes:
    (a) Bread rolls: knead the dough into circular or sausage-shaped rolls.
    (b) Braids: divide the dough into three or four equal parts. Roll each into a long strip. Damp the edges, join together and braid them.
    (c) Round loaves: use a round tin, such as a margarine or kimbo tin.

Maizemeal Bread
Asenath Sigot

Serves 8

Ingredients

1 cup maizemeal
1 cup baking flour
2 tbsps sugar
3 tsps baking powder
½ tsp salt
1 egg, beaten
1 cup milk
½ cup soft margarine

Method

1. Sift the maize flour, baking flour, sugar, baking powder and salt into a bowl.
2. Add the egg, milk and margarine. Beat for 2 minutes or until smooth.
3. Bake in an 8" square baking tin in a pre-heated oven (425 °F, 220 °C, gas mark 7) for 20–25 minutes.

Banana Bread
Sylvian Murungi

Serves 4

Ingredients

2½ cups flour
3 tsps baking powder
½ tsp salt
1 cup sugar
¼ cup margarine
1 egg, beaten

3 medium-sized ripe bananas, mashed
1 tbsp grated lemon rind
½ cup milk

Method

1. Preheat the oven to 350 °F (180 °C, gas mark 4). Grease a 9" × 5" loaf tin.
2. Sift the flour with the baking powder and salt.
3. Using a wooden spoon, beat together the sugar, margarine and egg until smooth.
4. Add the bananas, lemon rind and milk, mixing well. Add the flour and beat until smooth.
5. Pour the batter into the loaf tin and bake for about 60 minutes or until a fork inserted in the centre comes out clean.
6. Remove from the tin and allow to cool on a wire rack. To serve, cut into thin slices.

Rice Bread

Princess Barlay

Serves 8

Ingredients

2 cups roughly ground rice
3 cups mashed ripe bananas
½ cup cooking oil
1 cup warm water
¼ cup sugar
½ tsp salt
½ tsp bicarbonate of soda
½ tsp ground nutmeg

Method

1. Mix all the ingredients together and stir well.
2. Pour into a greased 8" loaf tin or round 8" cake tin and bake at 350 °F (180 °C, gas mark 4) for 30 minutes.

Cakes and biscuits

A cake may be served as a dessert: one major advantage is that it can be prepared ahead of the meal. The major ingredients of cakes are flour, eggs, sugar and margarine. A wide variety of cakes may be made by using different methods of preparation and by adding different flours, fruit, spices, herbs and seasonings.

The three basic methods of preparing cake batter are creaming, folding and rubbing or cutting in. The main skill is to incorporate air into the mixture. The recipe should be accurately followed and the ingredients used in the correct proportions.

Cakes should be baked in a moderately hot oven. When the cake is baking, the oven door should not be opened until about 15 to 20 minutes before the cake should be ready. The cake is ready when it attains a golden brown colour and when a toothpick or knife pricked into the middle comes out dry. When the cake is ready, it should be allowed to cool in the tin for 5–10 minutes, then put on a cake rack to finish cooling.

Home-made Baking Powder *Ruth Oniang'o*

Ingredients

$\frac{1}{4}$ tsp bicarbonate of soda
$\frac{1}{2}$ tsp cream of tartar
$\frac{1}{4}$ tsp starch
Mix well and place in an airtight glass container.

Groundnut Cake *Mary Makokha*

Ingredients

1 cup margarine
1 cup sugar
3 large eggs
1 cup pounded groundnuts
1 tbsp grated lemon rind
2 cups plain flour
2 tsps baking powder
$\frac{1}{4}$ cup milk

Method

1. Cream the fat and sugar together until fluffy.
2. Beat in the eggs, one at a time.
3. Stir in the groundnuts and lemon rind.
4. Sift together the flour and baking powder. Fold half into the mixture, add the milk, then fold in the remaining flour.
5. Put the mixture into a greased baking tin and bake at 330 °F (160 °C, gas mark 3) for $1\frac{1}{4}$ hours.

Banana Cake
Ruth Mititi

Ingredients

1 cup butter
1 cup sugar
juice and rind of 1 lemon
2 eggs
2 cups plain flour
2 tbsps cornflour
1 tsp baking powder
a pinch each of salt, ground ginger, cinnamon, cloves and nutmeg
4 bananas

Method

1. Cream the butter and sugar until light and fluffy.
2. Beat in the lemon rind, eggs, $1\frac{1}{2}$ cups of the flour and the remaining dry ingredients. If the mixture is too dry to combine, add a little milk.
3. Pour half of the mixture into a greased 9" cake tin.
4. Bake in a moderate oven (350 °F, 180 °C, gas mark 4) for 15 minutes. Allow to cool.
5. Peel and slice the bananas, sprinkle with lemon juice, then spoon into the tin and spread out.
6. Mix the remaining cake mixture with the remaining flour and pour over the bananas. Bake for 35 minutes.

Maizemeal Cake
Eunice Kirimi

Ingredients

1 cup maize flour
1 cup wheat flour
1 tsp baking powder
pinch of salt
$\frac{1}{2}$ cup sugar
pinch of grated lemon rind
$\frac{1}{2}$ cup margarine

2 eggs
1 cup milk
1 drop vanilla essence

Method

1. Sieve together all the dry ingredients and rub in the margarine.
2. Add the beaten eggs, milk and vanilla essence.
3. Pour the mixture into a greased tin.
4. Bake in a preheated oven at 375 °F (190 °C, gas mark 5) for 15 minutes.

Sweet Potato Cake

Florence Gimoi

Ingredients

$1\frac{1}{2}$ cups plain flour
2 tsps baking powder
$\frac{1}{2}$ cup margarine
$\frac{1}{2}$ cup sugar
2 large eggs
1 tbsp lemon rind
$1\frac{2}{3}$ cups milk
1 cup mashed sweet potatoes

Method

1. Sieve together the flour and baking powder.
2. Cream the fat and sugar until fluffy.
3. Beat in the eggs, one at a time.
4. Stir in the mashed sweet potatoes and lemon rind.
5. Add half the flour to the mixture, with the milk. Stir and fold in the remaining flour.
6. Pour the mixture into a greased baking tin and bake at 345 °F (180 °C, gas mark 4) for 1 hour.

Sorghum Cake

Elizabeth Nafula Kuria

Ingredients

$\frac{1}{2}$ cup margarine
$\frac{1}{2}$ cup sugar
2 eggs
2 tbsps milk
1 cup wheat flour
1 cup sorghum flour
2 tsps baking powder
2 tsps vanilla essence

Method

1. Cream the margarine and sugar until fluffy.
2. Add the eggs and beat well.
3. Add the milk and mix thoroughly.
4. Sift together the wheat flour, sorghum flour and baking powder, and add to the mixture. Add the vanilla essence.
5. Pour into a greased tin and bake in a moderately hot oven (375 °F, 190 °C, gas mark 4–5) for 35 minutes.

Cassava Cake
Damari Otieno

Ingredients

$1\frac{1}{4}$ cups wheat flour
$1\frac{1}{2}$ cups cassava flour
pinch of salt
4 tsps baking powder
$\frac{2}{3}$ cup margarine
$\frac{2}{3}$ cup sugar
3 eggs, beaten
1 cup milk
1 tsp vanilla essence
jam or butter icing for the filling

Method

1. Grease two medium-sized baking tins and dust with flour.
2. Sieve together the wheat and cassava flours, salt and baking powder.
3. Cream the margarine and sugar together until fluffy.
4. Add the beaten eggs a little at a time, beating well after each addition.
5. Add the flour a little at a time, folding in after each addition. Add the milk a third at a time, after each addition of flour. Stir in the vanilla essence.
6. Divide the mixture in half, put each half into a greased baking tin and bake in a preheated oven at 350 °F (180 °C, gas mark 4) for 30 minutes.
7. Cool the cakes and sandwich them together with plum jam or butter icing.

Variation: raisins or other dried fruit may be added with the vanilla essence.

Coconut Cake
Violet Mugalavai

Ingredients

2 tbsps margarine
1 cup caster sugar
4 eggs
$\frac{1}{4}$ cup milk
4 heaped tbsps dessicated coconut
2 cup plain flour
$1\frac{1}{2}$ tsps baking powder

Method

1. Cream the margarine and sugar until fluffy.
2. Whisk the eggs and add to the creamed mixture. Add the milk.
3. Fold in half of the dessicated coconut.
4. Sieve the flour and baking powder and fold into the mixture.
5. Pour the mixture into a prepared tin and bake at 375 °F (190 °C, gas mark 5) for 45 minutes or until a skewer inserted in the centre comes out clean.
6. Serve sprinkled with the rest of the dessicated coconut.

Cinnamon Cake
Dolrosa Ouma

Ingredients

$\frac{3}{4}$ cup margarine
$\frac{3}{4}$ cup sugar
2 cups plain flour
$1\frac{1}{2}$ tsps ground cinnamon
$1\frac{1}{2}$ tsps baking powder
6 tbsps milk
$1\frac{1}{2}$ tsps vanilla essence
2 eggs

Method

1. Cream the fat and sugar until fluffy.
2. Sieve the flour, cinnamon and baking powder together.
3. Mix the vanilla essence into the creamed mixture.
4. Beat the eggs one at a time and stir into the mixture.
5. Fold in the flour and add milk a little at a time until all the flour is folded in.
6. Pour the mixture into a greased tin and bake for 45 minutes at 375 °F (190 °C, gas mark 5).

Currant Cake
Cecilia Nderitu

Ingredients

2 cups plain flour
2 tsps baking powder
1 tsp salt
$\frac{3}{4}$ cup margarine
3 tbsps currants
1 cup sugar
1 egg
$\frac{3}{4}$ cup milk

Method

1. Sieve the flour, baking powder and salt together. Rub in the fat and add the currants and sugar.
2. Beat the egg and add the milk to it. Add this to the flour.
3. Mix with a wooden spoon to a soft dropping consistency. Pour into a greased tin and smooth the surface.
4. Bake in the middle of a preheated oven at 370 °F (190 °C, gas mark 5) for 15 minutes. Reduce the temperature to 350 °F (180 °C, gas mark 4). Bake for 30 minutes or until brown.

Orange Cake
Rose Osungu

Ingredients

1 cup margarine
1 cup sugar
3 eggs, beaten
grated rind and juice of $\frac{1}{2}$ orange
3 cups plain flour
2 tsps baking powder
pinch of salt
4 tbsps milk

Method

1. Cream the margarine and sugar together until fluffy.
2. Gradually add the beaten eggs and orange rind and juice.
3. Sieve together the flour, baking powder and salt and add, alternating with the milk, to the mixture in the bowl. Mix thoroughly and turn into a greased cake tin.
4. Bake in a preheated oven at 350–375 °F (180–190 °C, gas mark 4–5) for 35–40 minutes.
5. Allow to cool on a rack. Decorate with slices of orange.

Chocolate Cake
Julia Gitobu

Ingredients

2 cups sugar
2 cups margarine
2 eggs, beaten
$2\frac{1}{2}$ cups flour
2 tsps bicarbonate of soda
1 tsp salt
$\frac{1}{2}$ cup cocoa
1 cup sour milk
1 tsp vanilla essence
1 cup boiling water

Method

1. Cream together the sugar and margarine until fluffy.
2. Add the eggs. Sift together the remaining dry ingredients.
3. Add milk a little at a time, alternating with the dry ingredients.
4. Stir in the vanilla essence and boiling water.
5. Bake in an 8" greased cake tin at 350 °F (180 °C, gas mark 4) for 45 minutes.

Cherry Cake
Anne Mwangi

Ingredients

1 cup margarine
1 cup sugar
4 cups plain flour
1 tsp baking powder
6 glacé cherries
3 eggs
1 tsp vanilla essence

Method

1. Cream the margarine and sugar together until fluffy.
2. Sieve the flour and baking powder together.
3. Roll the cherries in flour and chop them into small pieces.
4. Beat the eggs and add to the creamed mixture, alternating with flour and using a metal spoon.
5. Add the vanilla essence and the chopped cherries. Mix thoroughly to a smooth consistency. Pour into a greased cake tin and bake at 325 °F (170 °C, gas mark 3–4) or until a skewer inserted in the centre comes out clean.

Yoghurt Cake
Diana Barwa

Ingredients

$\frac{2}{3}$ cup plain yoghurt
1 cup oil
1 cup milk
2 tbsps golden syrup
$\frac{1}{2}$ cup sugar
2 cups white flour
2 tsps baking powder
2 tsps bicarbonate of soda
2 cups Atta flour
To decorate:
2 tbsps orange marmalade
$\frac{1}{4}$ cup roasted groundnuts

Method

1. Mix together the yoghurt, oil, milk, golden syrup and sugar.
2. Sieve the flour with baking powder and bicarbonate of soda and fold into the liquid mixture. Stir until smooth.
3. Pour the mixture into a greased tin and bake for 1 hour at 300–350 °F (100–150 °C, gas mark 3) or until the cake has begun to separate from the edge and is brown on top.
4. Leave the cake to cool and then turn it out of the tin and spread the marmalade on top. Peel the groundnuts and decorate the cake with them.

Muffins
Magdalen Musila

Makes 8 muffins

Ingredients

$1\frac{3}{4}$ cups flour
$2\frac{1}{2}$ tsps baking powder
$\frac{3}{4}$ tsp salt
$\frac{1}{4}$ cup sugar
$\frac{3}{4}$ cup milk
1 egg
$\frac{1}{3}$ cup oil

Method

1. Preheat the oven to 425 °F (220 °C, gas mark 7).
2. Lightly grease the bottom of 8 muffin cups.
3. Sift together the flour, baking powder, salt and sugar. Blend together the milk, egg and oil.
4. Make a well in the centre of the dry ingredients and add the liquid, stirring gently.
5. Fill the greased muffin cups two-thirds full with batter. Bake for 20–25 minutes or until golden brown.

Mandazi
Dinah Makokha

(Doughnuts)

Serves 8

Ingredients

$2\frac{1}{2}$ cups flour
2 tsp baking powder
2 tsps sugar
1 egg
$\frac{1}{2}$ cup milk
oil for deep-frying

Method

1. Sieve the dry ingredients together and make a well in the middle.
2. Beat the egg and pour into the dry ingredients. Add the milk gradually and mix into the flour. Cover with a cloth and leave in a warm place for 10 minutes.
3. Heat the oil in a deep pan.
4. Meanwhile, roll the dough on a floured board to a thickness of $\frac{1}{2}$ cm.
5. Cut into small pieces and fry in hot fat until brown. Drain and serve hot with tea.

Note: mandazi originated from the Coast.

Sour milk Mandazi *Ruth Oniang'o*

Serves 8

Ingredients

2 cups flour
$\frac{1}{2}$ tsp bicarbonate of soda
$\frac{1}{2}$ tsp baking powder
$\frac{1}{4}$ cup sugar
$\frac{1}{4}$ tsp salt
1 egg
1 cup sour milk

Method

1. Sift together the dry ingredients and mix thoroughly.
2. Beat the egg and milk together.
3. Make a well in the dry ingredients and gradually add the liquid, mixing well to a thick consistency. Leave in a warm place for 10 minutes, then deep-fry, using a tablespoon to drop the mixture into the hot fat.

Variation: reduce the amount of sugar by half and fold one mashed ripe banana into the mixture at step 3.

Coconut Mandazi *Sarah Chilumo*
(Mahamris)

Makes 11–12 mahamris

Ingredients

2 tsps pressed yeast
2 tbsps sugar
$\frac{1}{2}$ cup coconut milk
$1\frac{1}{3}$ cups wheat flour
fat for deep-frying

Method

1. Mix the yeast with a little lukewarm water and 1 tsp of the sugar and leave for 15 minutes.
2. Sieve the dry ingredients together.
3. When the yeast starts to froth, make a well in the dry ingredients and pour in the yeast.
4. Add the coconut milk and mix well until the dough is firm.
5. Knead well and leave to prove in a warm oven for 45 minutes.
6. Knead the dough again, roll out to a thickness of $\frac{1}{2}$ cm and cut into circles or triangles. Leave for 15 minutes on a greased baking sheet.
7. Deep-fry in hot fat until golden brown.

Routis

Alice Kirigia

Makes 15 routis

Ingredients

$1\frac{1}{2}$ cups Atta flour
$\frac{1}{4}$ tsp salt
$\frac{1}{4}$ tsp pepper
$\frac{1}{2}$ cup milk
oil for deep-frying

Method

1. Sieve the flour, salt and pepper together. Make a well in the centre, add the milk a little at a time and mix with the flour, using a wooden spoon.
2. Knead and roll out to a thickness of $\frac{1}{2}$ cm with a rolling pin on a floured board. Cut into circles.
3. Heat the oil, checking the temperature by dropping a small piece of dough into it.
4. Drop the routis one at a time into the hot oil and cook for 1–2 minutes or until golden brown.
5. Drain on absorbent paper or a wire rack. Serve with stew and green vegetables.

Drop Scones

Janet Ngoci

Makes 16 scones

Ingredients

2 cups flour
$\frac{1}{3}$ cup sugar
$2\frac{1}{2}$ tsps baking powder

1 egg, beaten
1 cup milk
½ cup kimbo

Method

1. Sift the flour with the sugar and baking powder.
2. Make a well in the flour and pour in the beaten egg and milk.
3. Fold together until the mixture is of a dropping consistency (i.e. the mixture should drop from the spoon with a slight turn of the hand).
4. Heat a little fat in a frying pan.
5. Using a tablespoon, drop the mixture into the fat, a tablespoon at a time. Fry each side until golden brown.

Yam Drop Scones
Mary Mbae

Makes 16 scones

Ingredients

1 yam
2 cups flour
2½ tbsps sugar
1 cup milk
2 eggs
5 tbsps fat

Method

1. Peel and grate the yam.
2. Add the flour and sugar and mix well.
3. Add the milk, a little at a time, and mix well.
4. Beat the eggs and add to the mixture. Mix with a wooden spoon until the mixture is of a dropping consistency.
5. Heat a little fat in a frying pan and drop a tablespoon of the mixture at a time into the hot fat. When one side is browned, turn and brown the other side. Continue until all the batter has been used. Serve hot.

Maize Drop Scones
Dorcas Male

Makes 12 scones

Ingredients

1 cup wheat flour
½ cup maize flour

4 tsps sugar
1 egg, beaten
2 tbsps margarine
½ cup milk
½ tsp lemon juice
½ tsp lemon rind
5 tbsps fat

Method

1. Sieve all the dry ingredients together. Add the beaten egg, margarine, milk, lemon juice and rind.
2. Mix well and drop spoonfuls of the mixture into hot fat in a heavy frying pan. Fry on both sides until golden brown. Keep covered until cool.

Sweet Potato Biscuits

Zipporah Kabugu

Makes 16 biscuits

Ingredients

⅔ cup margarine
1 cooked and mashed sweet potato
¼ cup milk
2 tbsps sugar
1 cup flour
1 tsp baking powder

Method

1. Melt the fat and mix with the mashed sweet potato. Stir in the milk and sugar.
2. Sieve the flour and baking powder together and gradually add to the mixture.
3. Turn out on to a floured board, knead lightly and roll out to about ½ cm thick. Cut out with a biscuit cutter.
4. Put the biscuits on a greased baking tray and bake in a moderate oven (350 °F, 180 °C, gas mark 4) for 15–20 minutes.

Maizemeal Biscuits

Purity Mukiria

Makes 16 biscuits

Ingredients

1 cup maize flour
1 cup wheat flour

1 tsp baking powder
¼ cup margarine
¼ cup sugar
1 egg

Method

1. Sieve the flour and baking powder together and rub in the margarine. Add the sugar and mix well.
2. Add the egg and beat into a stiff dough.
3. Roll out on to a floured board and cut into shapes.
4. Bake in a moderately hot oven (400 °F, 200 °C, gas mark 6) for 20–25 minutes.

Simsim Biscuits

Agnes Masika

Makes 10 biscuits

Ingredients

1 cup simsim
3 tbsps margarine
1⅓ cups wheat flour
3 tbsps sugar
⅔ cup milk

Method

1. Roast and pound the simsim.
2. Rub the fat into the flour and stir in the sugar and simsim.
3. Add milk, little by little, to make a stiff dough.
4. Roll out to ½ cm thick, cut into shapes with a biscuit cutter, and place on a greased baking tray.
5. Bake in a moderately hot oven (375 °F, 190 °C, gas mark 5) for 10 minutes or until pale brown in colour. Remove from the oven and cool on a rack.

Baking Powder Biscuits

Margaret Maina

Makes 16 biscuits

Ingredients

2 cups flour
½ cup kimbo
2 tbsps sugar
4 tsps baking powder

½ tsp salt
½ cup milk
icing sugar to decorate

Method

1 Cut the kimbo into the flour, using a pastry blender, until you have a rich mixture that sticks together when pinched.
2 Add sugar, baking powder and salt, and enough milk to make the dough stick together.
3 Roll out and cut out with a biscuit cutter.
4 Place on a baking sheet and bake at 400 °F (200 °C, gas mark 6) for 15 minutes. Sprinkle with icing sugar.

Filled Butter Cookies
Leah Marangu

Makes 16 biscuits

Ingredients

1 cup butter
½ cup icing sugar
1 tsp almond or vanilla essence
2 cups flour
½ tsp salt

Method

1 Cream the butter and icing sugar until fluffy.
2 Add the almond or vanilla essence.
3 Sift together the flour and salt.
4 Gradually add the flour to the creamed mixture. Shape into small balls and place on a cookie sheet.
5 Press thumb gently into the centre of each ball. Bake at 375 °F (190 °C, gas mark 5) for 10–12 minutes.
6 Fill with any flavoured filling, preserves or jelly.

Custard Cream Biscuits
Rosemary Mutunga

Makes 8 biscuits

Ingredients

2 tbsps icing sugar
⅓ cup margarine

⅔ cup flour
4 tbsps custard powder
glacé cherries to decorate

Method

1. Preheat the oven to 400 °F (200 °C, gas mark 6). Lightly grease a baking tray.
2. Cream together the sugar and margarine, then stir in the flour and custard powder.
3. Mix with the hands until the mixture forms a smooth dough.
4. Turn on to a lightly floured board, roll out to a thickness of ½ cm and cut out into circles 4 cm in diameter. Press the top of each biscuit with a fork. Bake for 15 minutes, or until golden brown, on the middle shelf of the oven. Decorate with glacé cherries.

Pastry

8 or 9-inch one-crust pie shell

Ingredients

1 cup plain flour
½ tsp salt
⅓ cup and 1 tbsp shortening or ⅓ cup lard
2–3 tbsps cold water

Method

1. Mix together the flour and salt and cut in the fat until the mixture looks like fine breadcrumbs. Mix in the water, one tablespoon at a time, until all the flour is moistened. Knead the pastry into a firm ball.
2. Leave it in a cool place, preferably a refrigerator, for at least 30 minutes.
3. Roll out on a lightly floured board to about 4 cm larger than the inverted pie dish. Ease the pastry into the dish. Trim away any excess pastry.
4. For a baked pie shell, prick the pastry with a fork, put a handful of beans in the centre and bake in a preheated oven at 475 °F (240 °C, gas mark 9). This method is known as 'baking blind'.
5. For a two-crust pie, use twice the quantities of ingredients and roll the pastry into two circles. Line the pie dish with the pastry and turn the filling into it. Cover with the other half of the pastry, pressing the edges together, and cut slits in the top to allow steam to escape.

Note: if self-raising flour is used instead of plain flour, omit the salt.

Desserts, sweets and puddings

A sweet or dessert can play an important role in the meal, sometimes supplying some of the nutrients lacking in the main meal. Apart from making a meal more complete, it is pleasant to follow a main meal with a course that has a completely different appearance, consistency and taste.

Choosing a dessert will depend on the main meal and its accompaniments. The consistency and taste should always differ from that of the main meal. For instance, a main meal of ugali should not be followed by a steamed pudding.

Some desserts and puddings can be prepared ahead of time. The choice of a dessert may depend on the money and time available. Fruit and milk-based desserts are easy to prepare and very nourishing.

Banana Pudding
Gertrude Ireri

Serves 4

Ingredients

1 cup margarine
$\frac{1}{2}$ cup sugar
rind and juice of 1 lemon
1 egg, beaten
3 ripe bananas
$2\frac{1}{2}$ cups flour
3 tsps baking powder
$\frac{1}{4}$ tsp salt
$\frac{3}{4}$ cup milk

Method

1. Cream the fat, sugar and lemon rind until light and fluffy.
2. Add the beaten egg a little at a time, beating well after each addition.
3. Add the mashed ripe bananas and lemon juice and mix well.
4. Sift all the other dry ingredients together.
5. Add the milk, a little at a time, to the banana mixture, alternating with small quantities of the flour mixture.
6. Pour into a greased baking tin.
7. Bake on the middle shelf of a preheated oven at 350 °F (180 °C, gas mark 4) for 30–35 minutes.
8. Cool for 10 minutes in the tin, then remove from the tin and put on a cooling rack. Serve with custard.

Sweet Potato Pudding
Sophie Kirorei

Serves 2

Ingredients

5 tbsps margarine
6 tbsps sugar
2 eggs
6 tbsps flour
2 medium-sized sweet potatoes, boiled and mashed

Method

1. Cream the margarine and sugar until light and fluffy. Separate the eggs and beat the whites until stiff.
2. Beat the egg yolks into the margarine and sugar and stir in the flour.
3. Stir in the mashed potatoes and mix lightly until the mixture is quite smooth.
4. Fold in the egg whites.
5. Pour the mixture into a greased baking tin.
6. Bake in a moderate oven (350 °F, 180 °C, gas mark 4) for about 40 minutes, or until golden brown. Serve with custard.

Rice Pudding
Ruth Oniang'o

Serves 4

Ingredients

$1\frac{1}{2}$ cups rice
2 eggs
1 cup milk
$\frac{1}{2}$ tsp vanilla essence
pinch of salt
$\frac{1}{2}$ tsp nutmeg
$\frac{1}{3}$ cup sugar
1 tsp margarine

Method

1. Boil the rice (it should give approximately 3 cups when cooked).
2. Beat the eggs lightly, add milk, vanilla, salt, nutmeg and sugar and mix thoroughly. Add to the rice and stir well.
3. Pour into a greased casserole, cover and stand it in a pan of water. Bake in a preheated oven at 350 °F (175–180 °C, gas mark 4–5) for 45 minutes. After 40 minutes, remove the lid to brown the top.

Pineapple Upside-down Cake

Ruth Oniang'o

Serves 4

Ingredients

½ fresh pineapple
¼ cup water
1 cup sugar
¾ cup margarine
1 tbsp grated lemon rind
½ tsp vanilla essence
2 cups plain flour
2 tsps baking powder
pinch of salt
2 eggs
½ cup milk

Method

1. Chop the pineapple and cook it with the water and ¼ cup sugar until most of the water has evaporated and the pineapple looks shiny.
2. Grease a baking tin, coat the bottom with a little sugar and arrange the pieces of pineapple in it, with any syrup remaining from the pineapple.
3. Cream the remaining sugar and margarine until light and fluffy. Mix in the lemon rind and vanilla essence and beat in the eggs.
4. Add the sifted dry ingredients, alternating with the milk, and mix to a smooth consistency.
5. Pour the mixture over the pineapple and bake at 350 °F (180 °C, gas mark 4) until brown and firm on top. Serve hot with custard.

Shortbread Dessert

Asenath Sigot

Serves 4

Ingredients

For the shortbread:
1 cup margarine
½ cup and 1 tbsp sugar
2 cups plain flour
For the fruit salad:
1 cup chopped pineapple
1 cup chopped pawpaw
1 cup chopped banana

Method

1. Cream the margarine and ½ cup sugar together. Gradually work in the flour.
2. Roll the dough out on to an ungreased baking sheet, keeping the dough ½ cm thick.
3. Sprinkle with 1 tbsp sugar and prick all over with a fork.
4. Bake at 300 °F (150 °C, gas mark 2) for 45 minutes or until light brown.
5. Cut into squares and allow to cool. When cool, store in tightly covered containers.
6. Prepare a fruit salad and serve the shortbread with fruit salad on top.

Coconut Pie
Leah Marangu

Serves 6

Ingredients

6 eggs, beaten
4 tsps lemon juice
⅓ cup coconut flakes
1 tsp vanilla essence
8 tbsps sugar
½ cup melted margarine
unbaked 8" pastry shell (see p. 148)

Method

1. Combine all ingredients and pour into the pastry shell.
2. Bake at 350 °F (180 °C, gas mark 4) for 40 to 45 minutes or until a knife inserted halfway between the edge and centre of the custard comes out clean. Allow to cool and serve with whipped cream (optional).

Pumpkin Pie
Asenath Sigot

Serves 6

Ingredients

2 cups cooked mashed pumpkin
1½ cups condensed milk
2 eggs
½ tsp salt
1 tsp ground cinnamon
½ tsp ground ginger
½ tsp ground nutmeg
unbaked 9" pastry shell (see p. 148)
nuts to decorate

Desserts, sweets and puddings

Method

1. Combine the mashed pumpkin, condensed milk, eggs, salt and spices; mix well and pour into the pastry shell.
2. Preheat the oven to 425 °F (220 °C, gas mark 7).
3. Bake for 15 minutes and reduce the heat to 350 °F (180 °C, gas mark 4). Continue baking for 35-40 minutes or until a knife inserted in the centre comes out clean.
4. Allow to cool and decorate with nuts if desired. Refrigerate any leftovers.

Lemon Cake Pie

Leah Marangu

Serves 6

Ingredients

1 cup milk
2 tbsps margarine
juice of $1\frac{1}{2}$ lemons
3 tbsps sugar
3 eggs
2 tbsps flour
unbaked 8" pastry shell (see p. 148)

Method

1. Mix together the milk and margarine. Separate the eggs. Cream together the lemon juice, sugar, egg yolks and flour and add to the milk mixture.
2. Stir in the egg whites. Pour the mixture into the chilled pastry shell.
3. Bake at 400 °F (200 °C, gas mark 6) for 30-40 minutes, or until a knife inserted halfway between the outside and centre of the pie comes out clean. Allow to cool on a rack for 15-30 minutes. Chill in a refrigerator.

Coconut and Pineapple Pie

Princess Barlay

Serves 6

Ingredients

$\frac{1}{2}$ cup sugar
2 tsps plain flour
2 eggs
1 cup mashed pineapple
1 cup fresh grated coconut
$\frac{1}{2}$ cup margarine
2 tbsps milk

½ tsp vanilla essence
½ cup margarine
unbaked 8" pastry shell

Method

1. Preheat the oven to 350 °F (180 °C, gas mark 4).
2. Cream together the sugar, flour and eggs.
3. Add the rest of the ingredients and stir thoroughly.
4. Pour the mixture into the unbaked pastry shell.
5. Bake for 8 minutes or until the top is slightly browned.
6. Reduce the heat to 300 °F (150 °C, gas mark 2) and continue baking for 45 minutes, or until the pie is firm in the centre. Allow to cool before serving.

Vanilla Ice Cream   *Leah Marangu*

Serves 9

Ingredients

2   eggs
1   cup sugar
2⅔   cups milk
2   ripe bananas, mashed
2   tbsps vanilla essence
2⅔   cups cream
½   tsp salt

Method

1. Beat the eggs until stiff.
2. Gradually add the sugar, beating until the mixture thickens.
3. Add the remaining ingredients and mix thoroughly.
4. Freeze in an ice cream freezer until half-frozen. Remove from the freezer and beat thoroughly. Return to the freezer and leave until set.

Variations: instead of the bananas, use similar quantities of other mashed fresh fruit such as strawberries.

Mixed Fruit Crumble   *Anne Maina*

Serves 4

Ingredients

1   small pineapple
1   ripe banana

1 cup water
1½ tbsps sugar
1 tbsp shortening
1 cup flour

Method

1. Clean, peel and cube the pineapple and peel and slice the banana.
2. Steam for 5 minutes with 1 cup water and ½ tbsp sugar. Strain off the liquid.
3. Arrange the fruit alternately in a pie dish, sprinkling extra sugar between the layers if desired.
4. To prepare the crumble, rub together the fat, 1 tbsp sugar and the flour.
5. Cover the fruit with the crumble.
6. Place the pie dish in a baking tray containing water to a depth of one-eighth of the dish.
7. Bake at 350 °F (180 °C, gas mark 5-6) until golden brown.

Fruit Salad
Margaret Muri

Serves 4

Ingredients

2 lemons
2 oranges
½ cup sugar
2 ripe bananas
¼ medium pawpaw
¼ medium pineapple

Method

1. Grate and squeeze the lemons and oranges.
2. Mix the juices and rinds and boil for 10 minutes with the sugar.
3. Chill the syrup in a refrigerator or a cool place.
4. Peel the bananas and cut into thin slices. Peel the pawpaw and pineapple and cut into small cubes.
5. Put the fruit into a bowl and pour the syrup over it.

Never-Fail Peanut Butter Fudge
Peter Marangu

Ingredients

1½ cups sugar
1 cup milk
pinch of salt

2 tbsps margarine
¾ cup peanut butter
1 tsp vanilla essence

Method

1. Grease a 9" square tin.
2. Put the sugar, milk and salt in a saucepan. Bring to the boil and boil for 10 minutes.
3. Add the margarine and continue boiling to the soft ball stage.
4. Remove from the heat and add the peanut butter and vanilla essence.
5. Beat until the mixture begins to set.
6. Pour into the prepared tin and leave to cool. Cut into pieces when cool.

Pineapple Jam
Ruth Oniang'o

Ingredients

3 cups sugar
1 lemon
4 cups chopped fresh pineapple

Method

1. Warm the sugar in the oven. Squeeze the lemon and put the juice to one side. Chop the lemon peel and pulp, and put it into a muslin bag.
2. Heat the pineapple in a heavy pan. When it is heated, add sugar and the bag of lemon peel.
3. Heat slowly, stirring continuously, until the sugar dissolves. Bring slowly to the boil and simmer until the contents of the pan are clear.
4. Add the lemon juice and stir well. Remove any scum that forms on the surface of the jam.
5. Test for jelling, then bottle as described below.

Note: any frothing can be controlled by adding a little fat to the cooking jam.

Test for jelling: as the syrup thickens, two large drops will form along the edge of the spoon, one on either side. When these two drops come together and fall as a single drop, the 'sheeting stage' has been reached. This makes a firm jelly. For somewhat softer jelly or syrup, cook only until it falls in two heavy drops from the spoon.

Bottling: wash bottles and tops thoroughly in warm soapy water, rinse and dry thoroughly. Put cold jam into cold bottles, and seal with melted candlewax before screwing on the tops. The wax excludes air and prevents moisture forming. To melt candlewax, put the wax in a metal container and put the container in boiling water. The melted wax will solidify as it is poured over the surface of the cooled jam.

Beverages

Beverages are non-alcoholic drinks, such as tea, coffee, cocoa, milk and fruit juices. All beverages are useful in supplying the diet with liquid.

Tea and coffee are popular breakfast drinks in most Kenyan homes, and it is customary to serve tea not only at mealtimes but also at any time a guest arrives. Tea contains caffeine, which is a stimulant, traces of vitamin B complex and tannin. The best way to make tea is to heat a teapot by pouring hot water into it, pour off the water and put the tea leaves into the pot, allowing one teaspoon per person and 'one for the pot'. Boil fresh water, add to the tea immediately it has boiled and allow to infuse for 3–4 minutes.

Coffee also contains caffeine and, like tea, it is an infusion, so it is made in a similar way. Allow about 1 tablespoon of ground coffee for two people. Heat the coffee pot or jug with hot water, empty out the water and measure the coffee into the jug. Add boiling water, about a third of the jugful at a time, stir and wait 3–4 minutes after each addition. By the time the jug is full, the coffee grounds will have settled at the bottom.

Cocoa is gaining popularity as a beverage and has more nutritive value than tea or coffee because it contains fat, starch and iron.

Fruit juice can be used as a starter or as an accompaniment to meals or snacks, and in diluted form for feeding infants. Fruit juice is a good source of vitamin C.

Lemonade
Elizabeth Karani

Ingredients

4 lemons
4 cups water
6 tbsps sugar
drop of vanilla essence (optional)

Method

1. Grate the lemons and boil the rind in a little water for 5–10 minutes.
2. Squeeze the juice from the lemons.
3. Strain the lemon rind and keep the liquid.
4. Add the sugar to the lemon rind liquid, with the lemon juice, cold water and vanilla essence.
5. Chill in a refrigerator.

Assorted Fruit Punch

Eva Munene

Ingredients

2 lemons
2 cups hot water
5 passion fruit
1 orange
¼ pineapple
1 cup sugar
1 tsp tea leaves
ginger ale soda or Sprite

Method

1. Wash all the fruit thoroughly.
2. Grate the rind of 1 lemon and put the rind in the hot water. Cover the bowl and leave to stand for 15 minutes.
3. Scoop out the flesh of the passion fruit into a bowl.
4. Squeeze the orange and lemon.
5. Liquidise the pineapple or crush it and pass through a sieve to obtain the juice.
6. Mix the juices together and add the hot water and sugar. Stir thoroughly until all the sugar has dissolved.
7. Boil a little water and add the tea leaves.
8. Sieve the black tea and add it to the juice. Serve chilled with ginger ale soda or Sprite.

Orange Drink

Irene Gitahi

Ingredients

5 oranges
3 cups water
sugar to taste
a few drops of lemon essence

Method

1. Clean the oranges thoroughly, grate them and boil the rind in the water for 5-10 minutes.
2. Squeeze the juice from the oranges.
3. Strain the orange rind. Add the sugar, orange juice and lemon essence to the strained liquid. Dilute if necessary. Chill and serve cold.

Mixed Fruit Drink

Anne Mureria

Ingredients

5 oranges
4 lemons
5 cups hot water
8 passion fruit
sugar to taste

Method

1. Clean and dry all the fruit.
2. Grate the oranges and lemons. Pour the hot water on to the rind, cover and leave to cool.
3. Squeeze the lemons and oranges. Scoop out the passion fruit flesh into a jug.
4. Mix the juices together and whisk well.
5. When the water is cool, add it to the juice and add sugar to taste.
6. Strain and chill the juice.

Passion Fruit Drink

Asenath Sigot

Ingredients

10 passion fruit
5 cups warm water
sugar to taste

Method

1. Cut the passion fruit in half and scoop out the flesh into a jug containing the warm water.
2. Whisk the mixture well.
3. Add sugar to taste.
4. Strain the juice and chill. Serve cold.

Variation: mix with pawpaw juice or mango juice.

Egg Drink

Asenath Sigot

Ingredients

1 egg
$\frac{3}{4}$ cup milk
2 tsps sugar
pinch of salt
3–4 drops vanilla essence
pinch of ground cinnamon or nutmeg (optional)

Method

1. Beat the egg well and add milk, sugar, salt and vanilla essence. Stir until all the sugar is dissolved.
2. Strain and sprinkle with nutmeg or cinnamon.

Variation: cocoa syrup may be added.

Spiced Tea *Monica Duya*

Ingredients

6 cups water
1 tsp cloves
2 cm cinnamon stick or $\frac{1}{2}$ tsp ground cinnamon
$2\frac{1}{2}$ tsps tea leaves
$\frac{3}{4}$ cup orange juice
2 tsps lemon juice
$\frac{1}{2}$ cup sugar

Method

1. Combine the water, cloves and cinnamon and bring to the boil. Add the tea leaves, cover and allow to steep for 5 minutes, then strain.
2. Bring the orange juice, lemon juice and sugar to the boil. Stir and add to the hot tea. Serve hot.

Cinnamon Tea *Lois Ayuma*

Ingredients

2 cups water
$2\frac{1}{4}$ cups milk
2 tsps tea leaves
2 tsps ground cinnamon
sugar to taste

Method

1. Bring the water and milk to the boil.
2. Add the tea leaves, remove from the heat and stir well.
3. Add cinnamon and sugar to taste. Sieve and serve hot.

Recipes from other countries

Matoke Katogo
(Uganda)

Joy Awori

Serves 4

Ingredients

1 cup dried kidney beans
7-9 cups water
salt to taste
2 tsps curry powder
1 tbsp ghee
6-8 matoke

Method

1. Soak the beans overnight, then boil until tender. Leave sufficient water to make the sauce.
2. Add salt, curry powder, ghee and the peeled matoke and simmer until they are tender. Serve hot.

Creole Pea or Bean Soup
(Guyana)

Waveney Olembo

Serves 4

Ingredients

For the soup:
2 cups dried peas or beans
½ kg beef
1 tbsp kimbo
3 large potatoes
1 large onion
3 tomatoes
salt and black pepper to taste

For the dumplings:
2 cups plain flour

2 tbsps margarine
2 tsps salt
3 tsps sugar
1 tsp baking powder
a little chopped parsley
a little milk and water

Method

1. Soak the peas or beans overnight. Boil until tender.
2. Cut the beef into small pieces and fry in the kimbo until brown.
3. Peel and chop the potatoes, onion and tomatoes.
4. When the beans are ready, strain and mash them.
5. Combine the beans, beef, potatoes, dumplings, chopped onions, parsley, tomatoes and seasonings. Simmer until the potatoes are cooked.

To make the dumplings:
1. Sift the flour and rub in the margarine.
2. Add the salt, sugar, baking powder and parsley.
3. Mix with enough water and milk to make a stiff dough.
4. Knead lightly and form into balls in the palms of the hands.

Coconut Rice
(Liberia)

Princess Barlay

Serves 2

Ingredients

1 cup coconut milk
250 g meat
salt and pepper to taste
1 large onion
1 large tomato
$\frac{1}{2}$ tbsp tomato purée
$\frac{1}{4}$ beef stock cube
$\frac{3}{4}$ cup rice
2 carrots, sliced
$\frac{1}{4}$ cup shelled peas
1 tbsp margarine

Method

1. Put the coconut milk into a saucepan, add the cubed meat, salt and black pepper. Bring to the boil and simmer until the meat is tender (about 60 minutes).
2. Add the chopped onion and tomato, tomato purée and beef stock cube. Continue simmering for 15 minutes.
3. Wash the rice and add to the other ingredients in the saucepan, with the

carrots and peas. If the mixture is too dry, add a little water. Bring to the boil. Add the margarine and cook on a low heat, stirring occasionally, until the rice is cooked.

Cowpea Pudding
Cecilia Darkoh
(Ghana)

Serves 2

Ingredients

½ cup dried cowpeas
¾ cup hot water
1 small onion
salt to taste
½ tsp pepper
2 tbsps cooking oil

Method

1. Soak the cowpeas overnight, remove the skins, and grind in a blender with the hot water until smooth.
2. Grind the onion and add to the cowpeas.
3. Blend to a soft consistency with the salt, pepper and oil.
4. Grease individual custard cups or a large pudding mould and pour in the mixture to about two-thirds of the depth of the container.
5. Cover tightly with foil or greaseproof paper.
6. Steam for 1–2 hours, or for 30 minutes in a pressure cooker. Serve hot.

Cassava Pone
Waveney Olembo
(Guyana)

Serves 6–8

Ingredients

1 medium cassava
1 cup grated coconut
2 tbsps butter or margarine
¾ cup sugar
½ tsp vanilla essence
¼ tsp ground cinnamon
¼ tsp black pepper

Method

1. Peel and grate the cassava and coconut and mix with the butter or margarine.
2. Add the sugar, vanilla essence, cinnamon and black pepper and enough water to bind stiffly.
3. Put the mixture into a greased pan to a depth of about 2 cm.
4. Bake in a moderately hot oven (375 °F, 190 °C, gas mark 5) until crisp and brown on top. Cut into squares before serving.

Meat Patties
(Guyana)

Waveney Olembo

Makes 12 patties, serves 4

Ingredients

For the filling:
250 g minced meat
1 medium onion
1 clove garlic
1 tbsp fat
$\frac{1}{4}$ cup shelled green peas
1 medium sweet pepper, chopped
1 tbsp tomato sauce
1 stick celery, chopped
1 medium carrot, chopped
salt and pepper to taste
1 tsp curry powder
$\frac{1}{2}$ tbsp flour

For the shortcrust pastry:
1 cup fat
2 cups flour
2–4 tbsps water
1 egg, beaten

Method

1. Fry the minced meat with the chopped onion and crushed garlic until brown.
2. Boil the peas separately until almost soft. Add to the meat with the other vegetables.
3. Add salt, pepper, curry powder and tomato sauce.
4. Simmer with a little water until the vegetables are soft but not overcooked. Thicken with the flour if necessary. Allow to cool.
5. Prepare the shortcrust pastry (see p. 148). Roll out thinly and cut into circles to fit into patty tins.
6. Spoon some of the meat and vegetable filling on to each circle of pastry and cover with another pastry round. Use a fork to press the edges together.
7. Prick the top with a fork and brush with beaten egg.
8. Bake in a moderate oven (350 °F, 180 °C, gas mark 4) for 15 minutes. Serve as a main dish with tomato salad.

Chicken and Peanut Stew
(Ghana)

Cecilia Darkoh

Serves 4

Ingredients

1 medium chicken
2 tbsps corn oil
2 cups water
1 tsp salt
$\frac{1}{4}$ tsp black pepper
$\frac{1}{4}$ tsp white pepper
$\frac{1}{4}$ tsp cayenne pepper
$\frac{1}{4}$ tsp paprika
pinch of ground cloves
$\frac{1}{4}$ tsp ground nutmeg
pinch of ground ginger
$\frac{1}{2}$ cup peanut butter
1 onion, chopped
1 tomato, chopped
1 clove garlic

Method

1. Joint the chicken and brown the pieces in hot oil. Add the water and seasonings.
2. Cover the pot and simmer until the meat is tender.
3. Add the peanut butter, mixed into a paste with a little water, the onion, tomato and crushed garlic. Simmer until the onion is tender. Serve with rice.

Akka or Kodse
(Ghana)

Cecilia Darkoh

Serves 3

Ingredients

$1\frac{1}{2}$ cups dried cowpeas
pepper and salt to taste
1 medium onion
1 tbsp roiko mchuzi mix (optional)
1 egg, beaten
fat for deep-frying

Method

1. Soak the cowpeas overnight. Remove their skins and grind them to a smooth paste (a blender may be used).
2. Put them into a mixing bowl and add pepper, salt, grated onion, mchuzi mix and the beaten egg. Add milk or water if necessary to make a smooth consistency.
3. Deep-fry the mixture, a dessertspoonful at a time, until golden brown.

Fried Sweet Potatoes
(Liberia)

Princess Barlay

Serves 4

Ingredients

2 sweet potatoes
pinch of salt
2 cups oil

Method

1. Peel the raw sweet potatoes and slice them thinly lengthwise. Sprinkle with salt.
2. Fry slowly in the oil until brown and tender. Turn over and brown the other side.
3. Drain off excess oil on absorbent paper. Serve hot.

Fried Ripe Bananas
(Ghana)

Cecilia Darkoh

Serves 3

Ingredients

3 ripe bananas
$\frac{1}{2}$ tsp salt
2 cups oil

Method

1. Peel and slice the bananas lengthwise. Season with salt.
2. Fry, turning once, until brown on both sides.
3. Drain on absorbent paper. Serve hot.

Recipes from other countries 167

Sweet Potato Greens
(Liberia)

Princess Barlay

Serves 8

Ingredients

14 bunches sweet potato greens
1 beef stock cube
1 cup hot water
1 tbsp oil
2 onions
500 g beef or chicken
2 dried tilapia
1 pod hot pepper (optional)
black pepper and salt to taste

Method

1. Wash and chop the greens.
2. Dissolve the beef stock cube in the hot water.
3. Heat the oil, add the greens and chopped onions, and fry until the greens wilt. Cook the beef or chicken in a little water until tender.
4. Wash and bone the fish. Cut into small pieces.
5. Add the fish, cooked beef, chopped hot pepper, beef stock, black pepper and salt to the greens. Cook until the water has evaporated. Serve with rice.

Arrowroot Cake
(Liberia)

Princess Barlay

Serves 10–12

Ingredients

1 cup arrowroot flour
$\frac{1}{2}$ cup flour
$1\frac{1}{2}$ tsps baking powder
$\frac{1}{4}$ tsp ground cinnamon
$\frac{1}{4}$ tsp mixed spice
$\frac{1}{2}$ tsp ground nutmeg
$\frac{1}{8}$ cup raisins
$\frac{1}{8}$ cup roasted groundnuts
$\frac{1}{2}$ cup sugar
$\frac{1}{2}$ cup margarine
1 egg

Method

1. Sieve together all the dry ingredients except the sugar.
2. Wash and dry the raisins. Break the peanuts into small pieces using a chopping board and a rolling pin. Preheat the oven to 350 °F (180 °C, gas mark 4).
3. Cream the sugar and margarine until fluffy. Beat the eggs and mix, little by little, with the creamed mixture. Add the flour little by little, alternating with the milk.
4. Add the raisins and peanuts. Pour into a greased and flour-dusted 6″ baking tin.
5. Bake for 15–20 minutes. Remove and cool on a rack.

Tatale
(Ghana)   *Cecilia Darkoh*

Serves 4

Ingredients

4 over-ripe bananas
1 small onion, grated
1 tsp salt
½ tsp black pepper
5 tbsps plain flour
½ cup milk
1 egg
¼ cup fat

Method

1. Peel and mash the bananas to a smooth consistency.
2. Add the grated onion, salt, pepper and flour and mix thoroughly.
3. Add the milk to obtain a thick pouring consistency. Add the beaten egg.
4. Shallow-fry until golden brown (about 5 minutes).

Mugoyo
(Uganda)   *Joy Awori*

Serves 4

Ingredients

2 cups dried beans
4 sweet potatoes
salt to taste
1 tbsp margarine or butter (optional)

Method

1. Soak the beans overnight. Boil them until tender.
2. Add the peeled sweet potatoes and salt and continue cooking until the potatoes are tender and there is no excess water. Remove the sweet potatoes and set aside.
3. Mash the beans. Add the sweet potatoes and continue mashing until smooth. Add margarine or butter. Serve with a meat sauce and a green vegetable.

Jollof Rice
(Liberia)

Princess Barlay

Serves 6

Ingredients

1 kg chicken, skinned and boned
500 g stewing steak
salt and black pepper to taste
½ cup oil
2 onions
4 large tomatoes
3 tbsps tomato purée
6 cups water
3 cups rice
1 beef stock cube
2 carrots, sliced
⅓ cup shelled peas
⅓ cup green beans

Method

1. Chop the chicken and beef and season with salt and black pepper. Fry in oil until brown. Remove from the pan and sauté the chopped onions and tomatoes in the same oil.
2. Add the beef, chicken and tomato purée with 1 cup of water. Simmer for 15 minutes. Wash the rice and add it to the mixture with 5 cups of warm water.
3. Season with salt and black pepper and add the stock cube and the rest of the vegetables.
4. As soon as the mixture starts to boil, stir and reduce the heat. Simmer for about 30 minutes, stirring occasionally, until the water has been absorbed and the rice is cooked.

Fried Okra
(Liberia)

Princess Barlay

Serves 4

Ingredients

500 g beef
500 g okra
2 medium tomatoes
1 green pepper
2 medium onions
$\frac{3}{4}$ cup Elianto oil
$\frac{1}{2}$ tsp black pepper
1 tsp salt
200 g bacon or ham

Method

1. Cube and boil the beef for 40 minutes. Reserve the water.
2. Wash and slice the okra, tomatoes and green pepper and finely chop the onions.
3. Heat the oil in a deep saucepan and add the okra, onions, black pepper, salt and tomatoes. Fry until all slime disappears.
4. Remove the fat from the bacon or ham and cut into small pieces.
5. Add the bacon or ham, cubed cooked beef and the beef stock cube mixed with the water the beef was cooked in. Simmer for 10 minutes. Serve with rice or ugali.

Fried Collards and Cabbage
(Liberia)

Princess Barlay

Serves 8

Ingredients

3 bunches chopped collards (sukumawiki)
1 medium cabbage
$\frac{1}{2}$ cup oil
1 pod red or green chilli pepper
2 medium onions, chopped
1 kg chicken, diced and fried
500 g cooked ham or beef
1 beef stock cube
2 cups water

Method

1 Wash the collards and cabbage and shred them finely.
2 Heat the oil in a deep saucepan. Saute the onions and add the cabbage, collards and chopped chilli pepper. Fry for 10 minutes.
3 Add the chicken, ham or beef and the stock cube dissolved in 2 cups of water. Cook slowly on a medium heat for 20 minutes.
4 Season to taste with black pepper and salt and serve on a bed of rice.

Glossary of cooking terms

Bake	To cook in an oven or oven-type appliance, using a covered or uncovered container. Cooking meat in an oven in an uncovered container is usually called roasting.
Batter	A mixture of flour and a liquid, often combined with other ingredients, of a consistency thin enough to pour from a spoon.
Beat	To make a mixture smooth by introducing air with a quick, regular motion that turns the mixture over and over. An egg beater, a wooden spoon or an electric mixer may be used.
Blanch	To parboil in boiling water or precook in steam very briefly. The process is used to inactivate enzymes and to shrink some food, such as vegetables, before canning, freezing or drying. Vegetables are blanched in boiling water or steam and fruit in boiling fruit juice, syrup, water or steam. The process is also used to help remove the skins from nuts and some fruits and vegetables.
Blend	To mix thoroughly two or more ingredients.
Boil	To heat water or another liquid to a temperature at which bubbles rise continuously, breaking on the surface. The boiling temperature of water at sea level is 212 °F, 100 °C.
Casserole	A covered dish in which food may be cooked and served, or food that is cooked in this way.
Coagulation	The change of a substance from a fluid to a semi-solid state: curdling.
Cream	To soften a fat such as shortening or butter using a fork or other utensil, either before or while mixing with another food such as sugar.
Cutting in	Cutting fat into flour with a knife or pastry blender until it is finely divided.
Dough	A mixture of flour and liquid, sometimes with other ingredients. Dough is soft enough to knead or roll, as in making bread, but too stiff to stir or pour.
Dredge	To cover or coat with flour or another fine substance such as breadcrumbs or sugar.
Fold	To combine by using two motions, one cutting

Glossary of cooking terms 173

	vertically through the mixture, the other turning the mixture over by sliding the implement across the bottom of the mixing bowl.
Fry	To cook in fat (see **sauté**). Cooking in a deep layer of fat is called deep-frying.
Grill	To cook by direct heat from above, or an appliance used for cooking in this way.
Grind	To reduce to small particles by cutting or crushing.
Knead	To manipulate with the hands, using a pressing motion, accompanied by folding and stretching, for example in making dough or pastry.
Mince	To cut or chop into very small pieces.
Mix	To combine ingredients in any way that effects distribution.
Muffin cups	A traylike utensil containing a number of individual straight-sided or tapered cups. Also known as a cupcake pan.
Omelette pan	A small frying pan suitable for making omelettes.
Peel	To remove the skin or rind, usually from fruit or vegetables.
Pressure cooker	An airtight container for cooking food at a high temperature under steam pressure. This method is much faster than conventional cooking. Pressure cookers are used for cooking less tender cuts of meat and poultry and for cooking maize, beans and some vegetables.
Prove	In baking, to allow time for dough to be lightened by the fermentation of yeast.
Roast	To cook, uncovered, in hot air. Meat is usually roasted in an oven or over charcoal. The term is also applied to food such as maize or potatoes cooked in hot ashes, over charcoal or on hot stones or metal.
Roux	Equal parts of fat and flour mixed together and used for thickening soup or sauces.
Rubbing in	Rubbing fat into flour with the fingers until the mixture looks like fine breadcrumbs.
Sauté	To fry in a small amount of fat.
Scald	To heat milk or other liquid to the stage, just below boiling point, when tiny bubbles form at the edges.
Simmer	To cook in a liquid just below boiling point. Bubbles form slowly and collapse below the surface.
Smooth dropping consistency	The consistency of batter and similar mixtures that can be poured smoothly from a spoon.
Solution	A uniform blend of a solvent (liquid) and a solute (such as salt) dissolved in it.
Stew	To simmer food, normally cut into pieces, in a liquid for a long period. This method of cooking tenderises tough meat.

Sufuria	A saucepan without handles
Toast	To brown by dry heat.
Whip	To beat rapidly to incorporate air and increase volume. Generally applied to cream, eggs and dishes containing gelatin.

Glossary of food terms

Amaranthus	a green leafy vegetable growing wild in Kenya. It has a bland flavour and is often mixed with bitter vegetables.
Apoth	slippery vegetable (Luo)
Atta	brown wheat flour
Deg akeyo	hairy, bitter vegetable (Luo)
Dhania	coriander
Greengrams	tiny green leguminous seeds
Irio	in Kikuyu, irio literally means food. It also refers to a potato/legume/maize mash, which is part of the Kikuyu staple diet.
Isandi	a Kamba dish made of millet and maize flour
Isyo	a Kamba dish of maize, legumes and green leafy vegetables
Jollof rice	a casserole of rice, chicken and beef. This dish is from Liberia and is common throughout West Africa.
Kaimati	fermented doughnuts
Kibaki	a type of green leafy vegetable
Kiganda	originating from Uganda
Kimbo	vegetable fat
Kitoweo	a Kamba dish of meat, potatoes, bananas and other vegetables mixed or mashed together
Kivwea	a green, leafy vegetable (Kamba)
Kunde	cowpea leaves (Swahili)
Likhubi	cowpea leaves (Luhya)
Lumonde	a Meru dish of meat, bananas, green peas and potatoes
Madio	a mixture of arrowroot, maize and green peas with coconut milk
Mahamri	doughnuts (the name comes from the Coast)
Mandazi	doughnuts
Masala	powdered mixed spices
Mataha	irio with pumpkin leaves and bananas
Mataha ma kibaki	irio with green leafy vegetables
Matoke	green cooking bananas
Matumbo	tripe
Mawere	millet (Kamba)
Maziwa lala	sour milk
Meru muthikore	a dish made with pounded maize (Meru)

Miraaru	a type of cooking banana
Mrenda	a slippery green leafy vegetable (Luo)
Muchui	a dish of mixed tubers and other vegetables
Mugoyo	a Ugandan dish made with dried beans and sweet potatoes
Mukenye	a mixture of sweet potatoes and legumes (Luhya)
Mulee	a dish made with pounded cowpea seeds
Murere	a slippery green leafy vegetable (Luhya)
Muthokoi	a dish of pounded maize (Kamba)
Muthura	a type of irio made with potatoes, cowpeas and sorghum grains
Mwiko	wooden spoon
Nduma	arrowroot
Ngege	tilapia (Luo) – a species of freshwater fish
Ngerima	stuffed goat's stomach
Ngunza matu	a dish made with a mixture of maize flour, vegetables and/or meat
Njahe	white-eyed black beans
Njahe cia muciairi	food for a lactating mother
Njenga	the remains of sifted roughly-ground maize
Njugu	pigeon peas or groundnuts
Nowe	a large white kidney bean
Nyama choma	roast meat
Nyani	a dish made with meat, bananas, green peas and potatoes (Meru)
Omboga	amaranthus leaves
Omena	tiny freshwater fish (Luo)
Pone	a starchy dish from Guyana
Roiko mchuzi	a savoury flavouring powder (brand name)
Routis	fried bread (Asian)
Rumonde	a dish for infants, made with potatoes and cooking bananas
Ruguru	a medicinal porridge made with wimbi flour and arrowroot leaves
Simsim	sesame: a tiny oil seed that can be used whole or in a paste. It is popular in Western Kenya and at the Coast.
Sitayani	a sauce made with toasted pounded beans (Luhya)
Sukumawiki	kale. Sukumawiki literally means 'to push the week' and this vegetable is supposed to 'push' the meals during the critical time towards the end of the week, before payday.
Terere	amaranthus leaves
Tilapia	a freshwater fish
Ugali	a thick porridge that is part of many Kenyans' staple diet. It can be made with a single type of flour, such as maizemeal, or with a combination of flours, such as millet and sorghum.
Unga baridi	very fine flour
Wimbi	finger millet

References

Anazonwu-Bello, J.N., *Food and Nutrition in Practice*, Macmillan, 1976
Carson, B., *How You Plan and Prepare Meals*, McGraw-Hill Book Company, 1968
Hammond, B., *Cooking Explained*, Longman, 1984
Koeune, E., *The African Housewife and Her Home*, Kenya Literature Bureau, 1983
McGrath, H., *All About Food*, Oxford University Press, 1982
Powers, M.A., *Feeding the Family*, East African Literature Bureau, 1966
Ricketts, E., *Food, Health and You*, MacMillan, 1966
Robinson, C.H., and Lawler, M.R., *Normal and Therapeutic Nutrition*, MacMillan, 1977
Welbourn, H.F., *Nutrition in Tropical Countries*, Oxford University Press, 1963
White, R.B., *You and Your Food*, Prentice-Hall International, 1976

Index

Akka 165
Apoth in groundnut sauce 119
Arrowroot:
   arrowroot cake 167
   arrowroot stew 50
   arrowroot stew with meat or fish 47
   nduma casserole 113
   nduma casserole with cheese 113
   savoury arrowroot balls 15
Assorted flour porridge 10
Assorted fruit punch 158
Avocado pears, stuffed 21

Baking powder 134
Bananas:
   banana bread 132
   banana pudding 149
   bananas in coconut 123
   fried ripe bananas 166
   mashed bananas 40, 123
   matoke katogo 161
   matoke sandwich 21
   matoke tripe 67
   mchanyato 40
   miraaru hotpot 38
   stewed bananas 34
Beans, see Pulses
Bean and sweet potato mix 104
Beef:
   beef and bean stew 49
   beef and vegetable special 36
   beef broth 24
   beef stew with roast bananas 46
   dried or smoked beef stew 45
   minced beef stew 46
   sukumawiki beef stew 48
Biscuits:
   baking powder 146
   custard cream 147
   filled butter cookies 147
   maizemeal 145
   simsim 146
   sweet potato 145
  Bread: 131-134
   banana 131
   basic 131
   maizemeal 132
   rice 133
Bufuke 98

Cabbage, fried 122
Cabbage, crunchy cooked 122
Cakes:
   banana 135
   cassava 137
   cherry 140
   chocolate 139
   cinnamon 138
   coconut 137
   currant 138
   groundnut 134
   maizemeal 135
   orange 139
   sorghum 136
   sweet potato 136
   yoghurt 140
Calcium 2
Carbohydrates 1
Cassava:
   cake 137
   cakes 20
   cheese balls 14
   pone 163
Chapatis:
   coconut milk 85
   dhania 84
   flaky 84
Cheese eggplant casserole 41
Chicken:
   chicken and peanut stew 165
   chicken brochette coastal style 72
   chicken paprika 71
   chicken special 70
   chicken stew 68
   Kenya chicken 71
   Pakistan chicken curry 73
   simmered chicken 70
   spiced chicken in yoghurt 69
Coconut:
   and pineapple pie 153
   mandazi 142
   pie 152
   rice 162
Cocoa 159
Coffee 159

Cowpea leaves:
  likhubi 118
  in goundnut sauce 119
  in mushroom sauce 118
  sauce 128
Cowpea pudding 163
Creole pea or bean soup 161
Curried meat and vegetable stew 53

Drop scones 143
  maize 144
  yam 144

Eggs 27-32
  boiled 27
  egg drink 159
  fried 27
  fried egg spinach 117
  poached 27
  Scotch eggs 31
  scrambled 27
Eva's meatballs 56

Farmer's pie 61
Fats 1
Fermented porridge 11
Fermented sorghum and wimbi porridge 12
Fibre 3
Firinda 34
Fish 74-80
Fishcakes 80
Fluorine 2
Fried collards and cabbage 170
Fried fish fillets 79
Fried fish fillets and groundnuts 79
Fried mataha 100
Fruit crumble 154
Fruit salad 155
Fudge, peanut butter 155

Goat ribs 55
Goat's meat ngerima 58
Goulash 49
Green beans, sautéed 123
Green maize porridge 12
Green peppers, stuffed 62
Groundnut cake 134
Groundnut sauce 127

Ice cream, vanilla 154
Iodine 3
Irio:
  five cup irio 94
  fried irio 97
  with green peas 97
  with pigeon peas 96
  with pumpkin leaves 97
Iron 2
Isandi 92

Isyo, traditional 101
Isyo sya nthooko 102

Jollof rice 169

Kaimati 20
Kidney bean mash 104
Kikuyu sausage (mutura) 64
Kitoweo 38
Kodse 165

Lemonade 157
Lemon cake pie 153
Liver:
  omelette 66
  sautéed 65
  spiced coconut liver 65
  with yoghurt 66
Lumonde 60

Madio 87
Maize and beans 90
Maize and beans in sauce 91
Maize and cowpeas 90
Mashed green maize and peas 86
Mataha, fried 100
Matoke, see Bananas
Matumbo mash 39
Mau meat cakes 59
Mchanyato 40
Meat and vegetable pie 60
Meatballs 56
Meat patties 164
Meru muthikore 89
Minerals 2
Mixed green vegetables 115
Mixed fruit crumble 154
Mixed fruit drink 159
Mucui 35
Muffins 141
Mugoyo 168
Mukenye 99
Mukoomo 85
Mulee 105
Mulee, improved 105
Muree mash 101
Murere mix 120
Muthokoi:
  fried 87
  with cowpeas 88
  with meat 87
Muthura 95
Muthura mix 99
Mutton ribs:
  grilled 59
  roast 58
Mutura (Kikuyu sausage) 64

Nduma, see Arrowroot
Never-fail peanut butter fudge 155

Ngunja gutu 37
Ngunja gutu with meat 37
Ngunza matu 124
Njahe, sweet 102
Njahe cia muciairi 103
Njenga, creamed 106
Njugu 89
Nutrition 1-3
Nyama choma 54
Nyama choma, marinated 55
Nyani 91
Nyoyo 103
Nyuk abak bel 9

Offal 64-67
Okra, fried 170
Omelettes:
   green maize 30
   liver 66
   mixed vegetable 31
Omena in coconut milk 77
One-pot dishes 33-53
Orange cake 139
Orange drink 156

Pakistan chicken curry 73
Pancakes:
   arrowroot 30
   carrot 29
   sour milk 29
   sweet potato 28
Passion fruit drink 159
Pastry 148
Planning meals 7
Potatoes:
   brown potato balls 17
   mashed potatoes 112
   potato delight 19
   potato-meat roll 62
   sausage potatoes 22
Poultry 68-73
Phosphorous 2
Pineapple upside-down cake 151
Pineapple jam 156
Prawns, Lamu coconut 80
Pulses 94-106
Pumpkin, nutty 121
Pumpkin leaves, fried 121
Pumpkin pie 152
Punch, assorted fruit 156

Rice:
   bread 133
   carrot 82
   coconut 162
   green peas and coconut 83
   jollof 169
   minced meat 92
   pilau 93
   pudding 150
   spiced 83
Routis 145
Ruguru porridge 13
Rumonde 35

Sauces: 126-130
   cowpea 128
   cowpea leaf 128
   groundnut 127
   simsim 129
   sitayani (bean sauce) 127
   syrup 130
Seafish with coconut and ginger 78
Shortbread dessert 151
Simsim biscuits 146
Simsim sauce 129
Soups 23-26
   beef broth 24
   bone 23
   brown vegetable 25
   cream of tomato 25
   sutek 26
   vegetable 24
Spaghetti and cheese 41
Steamed vegetables 116
Stews 42-53
   arrowroot 50
   arrowroot with meat or fish 47
   bean 42
   beef and bean 49
   beef stew with roast bananas 46
   brown meat stew 43
   chicken 68
   curried meat and vegetable 53
   green banana 50
   meat 42
   meat and cabbage 47
   minced beef 46
   mixed meat 43
   mutton 45
   ngege 78
   pumpkin 52
   spicy meat 44
   stewed bananas 34
   stewed mushrooms 52
   yam 51
Storing food 5
Sukumawiki, fried 116
Sukumawiki in groundnut sauce 117
Sweet and sour meatballs 56
Sweet and sour pork 57
Sweet potatoes:
   baked in margarine 108
   balls 15, 112
   biscuits 145
   burgers 18
   cake 136
   cakes 18
   crisps 19

    fried 166
    glazed 107
    grilled 108
    mashed 110
    mashed enriched 111
    mukenye sweet potato rolls 109
    pancakes 28
    pudding 150
    spiced 17
    sweet potato and greengram mash 109
    sweet potato greens 167
    sweet potato meatballs 16
    sweet potato sausages 14
    sweet potatoes with meat 111
Syrup 130

Tatale 168
Tea 157
  cinnamon 160
  spiced 160
Tente 96

Terere dish 120
Tilapia:
  in coconut milk 74
  in mrenda 75
  in peanut sauce 76
  spicy tilapia 77
  stewed tilapia 75
Tubers 107–125

Ugali:
  enriched 82
  wimbi and maize 81

Vitamins 2

Wimbi porridge 10
  with sour milk 11

Yam:
  porridge 13
  potage 33
  stew 31